Depositional History of
Franchthi Cave

Excavations at Franchthi Cave, Greece

Fascicle 12

Depositional History of Franchthi Cave: Stratigraphy, Sedimentology, and Chronology

WILLIAM R. FARRAND

With a Report on the Background of the Franchthi Project by
Thomas W. Jacobsen

INDIANA UNIVERSITY PRESS

Bloomington & Indianapolis

Manufactured in the United States of America

This book has been supported by a grant from the National Endowment for the
Humanities, an independent federal agency

Cataloging-in-Publication Data

Farrand, William R.
Depositional History of the Franchthi Cave—
Stratigraphy, Sedimentology, and Chronology /
William R. Farrand, with a report on the
background of the Franchthi Project by
Thomas W. Jacobson.

(Excavations at Franchthi Cave, Greece; fasc. 12)
References: p.
1. Franchthi Cave Site (Greece). 2. Stone Age—Greece.
3. Excavations (Archaeology)—Greece. 4. Greece—Antiquities.
I. Farrand, William R.   II. Title.   III. Series

ISBN 0-253-21314-2

1 2 3 4 5 04 03 02 01 00 99

CONTENTS

FIGURES

TABLES

PLATES

PREFACE AND ACKNOWLEDGMENTS

It has been a long road since the day in summer 1974 when I agreed to take on the study of the sediments in Franchthi Cave. Tom Jacobsen invited me to the site because I had made similar studies on a number of prehistoric cave sites in the Near East and France. On my brief visit in 1974, I quickly realized that Franchthi was a very interesting cave with considerable time depth. It presented me with some challenges and opportunities that I had not experienced previously. I agreed on the spot to return for the next season—in 1976—to sample the sediments. The 1976 season was to be the final digging season in Franchthi. All the trenches inside the cave were still open, although some were somewhat slumped or not very clean. The disadvantage of joining the project so late, of course, was not being at the site during the preceding seven seasons, during which much sediment was removed. This meant that I had to rely on field notebook descriptions and drawings of sections long disappeared by excavators who were no longer associated with the project.

So in 1976 the sediments were sampled selectively; all the sections were examined, as well as the surrounding landscape. I returned during the study season in 1979 to work with the project draftsman, Dan Pullen, on the preparation of the final section drawings and plans. We compared all the extant sections against the drawings that had been made over the eight years of field work. It turned out that my background in geology, and in cave stratigraphy in particular, earned me the *de facto* assignment of cave stratigrapher for the project. This was not a job that I had counted on, but it is one of prime importance on any stratified site. It was a great opportunity for me, for which I am grateful, not only concerning Franchthi, but in theoretical terms as well.

It is important to say that this fascicle does not stand by itself. There are many cross-references herein to the contributions of my Franchthi colleagues, to which the reader should refer for elaboration of their ideas and their data.

The project team had two fruitful opportunities to meet together for symposia in Bloomington, Indiana, to discuss various aspects of phasing and overall stratigraphy (more details in Chapter 2). Then, I was able to spend the Fall of 1985 in Bloomington studying the notebooks and drawings. Unfortunately I was sidetracked by academic duties thereafter, and Franchthi was put on the shelf until recently.

My initial approach to the sediment study was somewhat circumscribed because I was coming to the project belatedly. However, once I became concerned with the stratigraphy and correlations in the cave, and with deciphering the notebooks, I wished that I had sampled more extensively and in all the cave trenches. It was too late at that point because the excavations were shut down. In any case, the excavation documents contained a great amount of information, and increasingly so as the excavation seasons proceeded. Nevertheless, some questions can never be completely answered from the available information.

I am very grateful to all my Franchthi colleagues for helping me to understand their work and all that had transpired before I arrived on the scene. Tom Jacobsen, of course, made this all possible with his foresight in bringing me into a well-organized, interdisciplinary project. His approach to a complex site such as Franchthi was exemplary, and he did a great job as master of ceremonies. His close and valued friendship is one of my most appreciated rewards from working at Franchthi. In addition, Tom has contributed valuable historical information on the site and on the project in general in the first and second chapters of this fascicle.

Kaddee Vitelli's leadership in these past few years has been a great boon, especially in seeing us through the publication stage. Moreover, she and I continue to have interesting discussions on stratigraphic matters. Catherine Perlès also has helped keep me honest in my interpretations and extrapolations. Tony Wilkinson showed me the landscape and shared his ideas on the geomorphology of the area around Franchthi during my first visit in 1974, and I greatly appreciate his insights into aspects of the cave stratigraphy, as well. Tjeerd van Andel has also become a valued colleague over the course of these years, and he and Julie Stein have been very helpful in reviewing the earlier version of this fascicle and in making very useful suggestions for improving it. Julie Hansen and I have worked closely on questions of stratigraphy, particularly biozonation. Nancy Whitney-Desautels' study of land snails has contributed very interesting, serendipitous data for interpreting the Franchthi depositional history, and she and I worked closely on approaches to land-snail biozonation for, I hope, our mutual benefit. Susan Duhon spent some time in Ann Arbor voluntarily helping with the laboratory analyses, along with Larry Davis, one of my Michigan master's degree students.

Kay Clahassey, artist in the University of Michigan Museum of Anthropology, has done a magnificent job and shown much patience in translating my rough drawings into handsome figures.

Finally, I am deeply indebted to Carola and Michelle, not only for their forbearance, but also for their encouragement while I have been working too long and too hard on completing this work.

Financial support for my Franchthi research has come through various grants to T. W. Jacobsen and to K. D. Vitelli through the Program in Classical Archaeology at Indiana University. I am also grateful to the Indiana University Institute for Advanced Studies for a fellowship during my research leave in Bloomington in 1985.

Depositional History of Franchthi Cave

CHAPTER ONE

Background of the Franchthi Project

Thomas W. Jacobsen

THE ARGOLID EXPLORATION PROJECT

The Franchthi excavations began as part of a larger undertaking known in its initial stages as the "Argolid Exploration Project" (AEP). This is something of a misnomer, because the project was primarily concerned with the investigation of only the southern portion of the Argive peninsula, an area of some 700 km² known as "Ermionis" or the "Hermionid."

(For a general description of the landscape and geographical names in this area, see van Andel and Sutton [1987:Chap. 1] or Jameson et al. [1994:Chap. 1]. Ancient Halieis, mentioned below, is exactly 4 km due south of Franchthi Cave.)

The AEP was the brainchild of M. H. Jameson, who, as a young postdoctoral scholar, first visited the area in 1950. During the course of the following decade, he conducted epigraphical and topographical research in the Hermionid while beginning to lay plans for the larger project. It was not until 1962 that systematic field work was initiated with the beginning of the excavations at ancient Halieis near Porto Kheli under the overall direction of Jameson and John H. Young. Additional field work was carried out at Halieis in 1965 and 1966, all in the name of the University of Pennsylvania and under the aegis of the Greek Archaeological Service and the American School of Classical Studies in Athens (cf. Merrit 1984:212 ff.). The writer, first as a graduate student at Pennsylvania and later as a young faculty member at Vanderbilt University, took part in those early excavations at Halieis, during which time considerable attention was also given to informal survey in the neighborhood of the ancient site. Indeed, it was in the course of such reconnaissance that we first visited the cave in the headland of Franchthi.

The AEP began to assume a more formalized structure during the academic year 1965–1966, when the writer was invited to become codirector of the project as a whole and field director of the Halieis excavations. The collaboration was solidified when Jacobsen moved to Indiana University in the autumn of 1966. The AEP then became a joint undertaking of the University of Pennsylvania and Indiana University, with Jameson and Jacobsen as codirectors of the overall project, the latter in charge of the land excavations at Halieis and the former directing the recently begun work underwater in the bay of Porto Kheli.

With the beginning of excavation at Franchthi Cave in 1967, the overall program of the AEP became more complex (see also Chapter 2). Excavation was also carried out at a nearby modern kiln site (Lorenzo), the underwater work at Halieis continued, and additional informal reconnaissance was conducted in the area. This diverse program of field work—excavation at Franchthi Cave, excavation at Halieis (land and underwater), and surface surveying—was maintained through the 1968 and 1969 field seasons under the collaborative direction of Jameson and Jacobsen, until it became apparent (spring 1969) that a restructuring of the project was necessary. At that point W. W. Rudolph was hired by Indiana University to become field director of the land excavations at Halieis, under the general direction of Jameson, who also continued to oversee the underwater work. Jacobsen, while continuing as codirector of AEP, was then free to concentrate on the excavations at Franchthi Cave.

In order to put the two excavated sites in their regional context, plans for a proper archaeological survey of the Hermionid began to be drawn up during the academic year 1970–1971. Funding and staff were secured, and the first season of systematic reconnaissance was carried out in the summer of 1972 under the direction, successively, of Jameson, Jacobsen, and James A. Dengate. Thereafter, given Jacobsen's increasing responsibilities for the completion of the Franchthi excavations and the organization and study of the excavated materials and other data for publication, further immediate work on the survey was postponed. It was not resumed until Jameson had moved from Pennsylvania to Stanford University in 1976. The "Archaeological and Environmental Survey of the Southern Argolid," the official successor of the AEP, began in 1979 under the direction of Jameson and Tjeerd H. van Andel and was completed in 1984. That project, incorporating the results of all previous AEP surveys, has now been published (Jameson et al. 1994).

PREHISTORIC STUDIES IN GREECE IN 1966

It is important to clarify at the outset the research orientation and the goals of this project, but, before doing so, the intellectual climate or setting in which the Franchthi excavations began needs to be made clear. What was the state of Aegean archaeology in the mid-1960s, and what did we know (or think we knew) about Greek prehistory in those days? What questions were we asking then? While it is neither appropriate nor my intention to offer an exhaustive reconstruction or evaluation of scholarship or research of that period, something needs to be said about it in order to appreciate what follows—especially in view of the rather dramatic advances in Aegean prehistoric studies since that time.

The logical place to begin, it seems to me, is with the late Saul Weinberg's comprehensive synthesis of the Aegean "Stone Age," which first appeared in fascicle form in 1965 (Weinberg 1965).[1] It was an excellent and well-documented overview of prehistoric (pre-Bronze Age) archaeology in Greece that appeared on the eve of our first season of excavation at Franchthi Cave. Weinberg's presentation clearly reflected the limits of our knowledge at the time and was groundbreaking in several ways. It was the first study in English to give serious attention to pre-Neolithic settlement in Greece, and it was the most comprehensive survey of the Greek Neolithic period since that of Mylonas (1928). More important in my view, Weinberg was among the first (after Childe) to consider Aegean prehistory in its larger European and eastern Mediterranean context.

Very little was known about the earliest stages of Greek prehistory in the early 1960s. Systematic investigation of the Palaeolithic was just beginning, and the important preliminary reports of the Cambridge University–British Academy project under the direction of the late E. S. Higgs (Dakaris et al. 1964; Higgs and Vita-Finzi 1966) were not yet available to Weinberg as he compiled his synthesis. Yet he was aware of their existence as well as, for example, of the ongoing work of the Germans in Thessaly and the French in Elis.

The "Mesolithic" (Weinberg used the term unabashedly) was essentially terra incognita.

There were, to be sure, reports of scattered surface discoveries of "microliths" (the implicit assumption always being that they belonged to the Mesolithic), but, apart from the tantalizing publications of Markovits from the 1920s, Weinberg could point to no stratified evidence of a Mesolithic facies or horizon in Greece. As it turned out, the first good evidence for that was only just then being exposed at the small coastal site of Sidari on the island of Corfu, but scientific reports had not yet been published. Nevertheless, it is characteristic of Weinberg's approach that, in attempting to make sense of the early work of Markovits, he concluded (Weinberg 1965:9), "it is clear that we are only at the beginning of the discovery of Palaeolithic and Mesolithic remains in Greece; it is also clear that while skepticism is healthy, negativism is antiprogressive."

The Neolithic of Greece was much better known in 1965, especially in the north (Thessaly), where rich assemblages had been well known since the turn of the century and where the Greeks and the Germans were continuing to conduct important excavations. But a particularly noteworthy departure from traditional practice in Aegean archaeology was the work of a Cambridge team at Nea Nikomedeia (Macedonia) in the early years of the decade. There, essentially for the first time in Aegean archaeology, attention was being given to the palaeoenvironmental context of a Neolithic site in Greece, and, as with the Germans in Thessaly, efforts were finally being made to gather information about Neolithic subsistence practices.

The Neolithic of southern Greece, noticeably different in many respects from that of the north, was rather less well understood. There had been a certain amount of excavation over the years, but most of the material had not been fully reported, or it lacked a solid stratigraphic basis. It is somewhat ironical, then, that the tripartite chronological scheme (EN-MN-LN), developed earlier by Weinberg and most forcefully proposed by him in 1965, was inspired largely by his work at one of those sites where clear stratigraphy was a problem, namely Old Corinth. While he recognized this scheme as "most pertinent to the Peloponnese" (Weinberg 1965:17), he applied it to the Greek mainland as a whole and cautiously concluded, "significant cultural changes are sufficient to make valid the tripartite scheme, and its use, in turn, makes easier the handling of the large body of material now at hand" (Weinberg 1965:18). Almost at once, it became the most widely used means of describing the Neolithic chronological sequence in Greece as a whole.

What were the burning questions and major unresolved problems of prehistoric research in Greece in the mid-1960s? Chronology, always a primary preoccupation (often exclusively so) with Aegean archaeologists, continued to be a hotly disputed issue. Weinberg's proposed scheme for the mainland Neolithic provided a new way of looking at the relative sequence, and, as a "believer" in the radiocarbon method, he even offered a general absolute framework based on the handful of measurements then available. Other archaeologists were less willing to accept this novel method, and therefore a corpus of radiocarbon dates was slow to accumulate in Greece. In any case, the pre-Neolithic was still "floating," and it was not until 1966 that the first radiocarbon dates from the Greek Palaeolithic were published (Higgs and Vita-Finzi 1966).

The problem of "origins" has also been an abiding concern of archaeologists working in Greece, and, given the new discoveries of early agricultural settlements at sites such as Knossos in Crete, Nea Nikomedeia, and "Aceramic" Argissa in Thessaly, it was no less so in the 1960s. Indeed, it was a particularly important issue because of the recent revelations about the "agricultural revolution" in the Middle East by Braidwood (preliminarily articulated in Braidwood 1960). Major questions for Weinberg (1965:14–15), then, were "whether or not the Aegean area played any part in the revolution that changed man from a food-gatherer to a food-producer" and "what was the relationship of the Aceramic Neolithic of Greece to the cultures which preceded and succeeded it?" He answered both questions, on the basis of available evidence and then-current explanatory devices, in essentially the same way:

Nothing yet found in Greece, or in the European countries to the north, suggests that a similar process of change took place in Europe independently. It is more likely that the inhabitants of Greece received from Anatolia or farther east the benefits of a revolution already accomplished, in this case chiefly a knowledge of agriculture and the raising of domesticated animals, permitting permanent settlement. . . . If, as seems likely, the Aegean received its settlers of the Aceramic Neolithic period from the Near East, then this was perhaps the first of a long series of westward movements into the Aegean, possibly already by boat. . . . We are thus left at present without any indication of relationship between the Mesolithic and Aceramic Neolithic cultures. . . . On the other hand, all [Greek Aceramic Neolithic] sites have Early Neolithic pottery phases immediately above the Aceramic stratum. . . . What seems most probable is that the makers of pottery took over these sites directly from their former occupants, so that there was a continuity of occupation, though very likely without a cultural continuity. At present, the first Ceramic Neolithic phase cannot be shown to have developed locally out of the Aceramic Neolithic culture. (Weinberg 1965:14–16)

Explanation of change in Aegean archaeology has commonly been attributed to events (rather than processes), such as migrations or invasions from elsewhere, most often the Near East, and *ex oriente lux* explanations of the Greek Neolithic dominated the thinking of the 1950s and the 1960s.

At the same time, in spite of the emerging interest in the environment and subsistence economy, artifact analysis (principally of ceramics and "architecture," to the neglect of virtually all other remains) still dominated the archaeologist's interests. Organic remains were only beginning to be considered (cf. Hansen 1991 and Payne 1985). This was reflected in the then-current methods of excavation and field techniques. Apart from the screening of smaller special deposits with hand shakers, sieving (whether wet or dry) was normally not practiced on excavations in Greece. Sampling strategies on excavations were often unclear, and, in those cases where a substantial portion of the sample was discarded (the majority of the excavations?), there were rarely explicit criteria for those decisions. Stratigraphy seems to have been regularly and carefully observed, but the frequent omission of published section drawings (owing to so many preliminary reports?) made it difficult for the reader to verify stratigraphic interpretations.

The decade of the 1960s was, in many respects, a watershed in the study of Aegean prehistory. While many of the traditional concerns, such as chronology and "cultural" affinities, continued to be of importance to archaeologists working in Greece, new questions were also being asked. As we have seen, the latter were largely in the realm of "palaeoeconomy" and "palaeoenvironmental reconstruction" and were much indebted to the pioneering work of scholars such as Braidwood, Childe, Grahame Clark, and Higgs. At the same time, there was much exciting work going on elsewhere in the eastern Mediterranean region (e.g., Hole et al. 1969; McBurney 1967; and Mellaart 1967), and it is significant that Aegean archaeologists were becoming aware of that work.

CONCEPTUAL ORIENTATION OF THE PROJECT

While most archaeologists today are deeply concerned with establishing and articulating coherent and explicit theoretical models or frameworks for their research, such was not the case in Aegean (Greek) archaeology in the 1960s. To be sure, there was often a genuine sense of problem, but even that was rarely made explicit. This is not to say, however, that most of us brought up in those days were working without a basic set of assumptions—implicit though they may have been and varied as they were from individual to individual—about archaeology, its goals, its potential, and its limitations. There were clearly schools of archaeological thought and philosophies of archaeology then as now, but they do not appear to have been so clearly defined or articulated as they are coming to be today.

My aim in this section, then, is to identify some of those assumptions and points of view, specifically those that contributed most significantly to the thinking that lay behind the formulation of this project. My classification of that thought is admittedly personalized and rather autobiographical, as inevitably it must be. While these points of view are certainly shared by most members of our project, they will vary in degree and really are unique to the writer. Accordingly, I have identified three such schools of thought, which I would characterize as follows.

Euro-American Classical Archaeology

Renfrew (1980) referred to classical archaeology as the "Great Tradition," but it is regrettable that the philosophical and theoretical underpinnings of this school have never been fully articulated. Classical archaeology in the broadest sense has included from very early on a significant prehistoric component, but its approach has always been primarily historical. The work of the most notable early Aegean prehistorians in America (e.g., Carl W. Blegen and his successor, J. L. Caskey) may be characterized as meticulously descriptive and aimed at reconstructing cultural history with cautious empiricism. The classical archaeologist has traditionally been most concerned with the recovery and classification of data and the establishment of cultural chronologies. Classical archaeology's traditionally close ties with classical philology have resulted in a natural affiliation with humanistic studies rather than with the social (or natural) sciences and have contributed to a tradition which, as Renfrew suggests, sets high priorities on careful scholarship and comprehensive publication. The traditional focus of classical archaeology was on the site and its "monuments" (man-made objects), and it was not until the late 1950s and 1960s, largely through the efforts of another of Blegen's disciples, W. A. McDonald, that serious attention began to be paid to regional studies and the palaeoenvironmental contexts of archaeological sites. The Franchthi excavations (as well as the Argolid Exploration Project) owe a great deal to the work of McDonald and his colleagues on the Minnesota Messenia Expedition (McDonald and Rapp 1972).[2]

British Prehistoric Archaeology

Prehistoric archaeology in Britain has traditionally been related to classical archaeology in its natural affinities with historical studies and the culture-historical approach. Unlike prehistoric archaeology in North America, British prehistoric archaeology is usually separated from anthropology and the social sciences. Moreover, it differs from American classical archaeology in several ways, not least in the larger worldview of its greatest practitioners, scholars such as V. G. Childe or J. G. D. Clark. Indeed, it is the work of Grahame Clark, as much as any individual, that influenced the early shaping of the Franchthi project. His seminal *Prehistoric Europe, The Economic Basis* (1952) and *Archaeology and Society* (3rd ed. 1957) drew attention for the first time, at least for this writer, to the various dimensions of prehistoric economies and the importance of multidisciplinary palaeoenvironmental studies to archaeology:

> What the prehistoric archaeologist has to study is the history of human settlement in relation to the history of the climate, topography, vegetation, and fauna of the territory in question. One of the greatest difficulties in such a study is to distinguish

between changes in the environment brought about by purely natural processes and those produced, whether intentionally or incidentally, by the activities of human society, and this can only be resolved by intimate co-operation in the field with climatologists, geologists, pedologists, botanists, zoologists, palaeontologists in the comradeship of Quaternary research. The archaeologist wants to know precisely what geographical conditions obtained at each stage of human settlement; the extent to which the economic activities of any particular community were limited by the external environment; and above all how far the economic activities of the people he is studying are reflected in and can be reconstructed from changes in the geographical surroundings. It is only by observing the human cultures of antiquity as elements in a changing ecological situation that it is possible to form a clear idea of even the economic basis of early settlement, see precisely how early man utilized his environment and so arrive at a fuller understanding of his intellectual, economic, and social progress. (Clark 1957:20)

There is little here that I would change if rewriting this statement today.

I should add, however, that the above point of view was reinforced and further refined by the important work carried out by Clark's Cambridge colleagues on the British Academy Major Research Project in the Early History of Agriculture under the direction of E. S. Higgs, a project that had gotten underway in northern Greece shortly before the outset of our excavations. That project's concern for "paleoeconomy" and "site-catchment analysis" developed the earlier views of Clark and, along with the important field work of other British prehistorians, Robert Rodden at Nea Nikomedeia and John Evans and Colin Renfrew at Saliagos, introduced the first systematic investigations of past man–land relations into archaeology in Greece. Our excavations have benefited greatly from interaction with all of those projects.

Finally, the work of yet another of Clark's Cambridge colleagues, the late C. B. M. McBurney, also had a noteworthy impact on our work. The excavations conducted by McBurney at the Haua Fteah in Libya were a milestone in cave archaeology in the eastern Mediterranean, and the publication of the results of those excavations (McBurney 1967) coincided with our first season of fieldwork at Franchthi. The Haua Fteah report therefore served as an early inspiration to us, especially in terms of the excavation of a cave site.

American Anthropology and Anthropological Archaeology

I include under this rubric the large field of anthropological studies in North America, of which archaeology is normally regarded as one of the major subfields. Anthropology has for some time firmly allied itself with the social sciences in American universities, and "anthropological archaeology" (e.g., Gibbon 1984, 1985) has tended to distance itself from the more historically oriented practice of archaeology associated with humanistic disciplines such as classical studies, art history, and Near Eastern studies. This ahistorical (or even antihistorical) and nonhumanistic approach to the study of the past, while still prevalent to some degree, may have reached a peak in the 1960s and 1970s (cf. Trigger 1986), the floruit of what Renfrew (1980, 1983) has called the "New Americanist Archaeology." Renfrew aptly characterized this dichotomy in the study of archaeology in America as the "Great Divide," and, under the circumstances, it is fair to say that anthropological archaeology (or, at least, the "new" archaeology) contributed little to the formation of the Franchthi project.

On the other hand, archaeology is practiced by individuals, and one of the most distinguished of the practitioners of "anthropologically oriented archaeology" (as he has put it) is R. J. Braidwood of the University of Chicago. The multidisciplinary approach of Braidwood and his team on the Iraq–Jarmo Prehistoric Project (Braidwood and Howe 1960) probably influenced more than any other single enterprise the practice of prehistoric archaeology in the eastern Mediterranean region in the last 40 years. Given the state of prehistoric studies in Greece in the mid-1960s and the impact of Braidwood's work on the questions then being asked (see above), it is not surprising that his approach had a significant effect on the development of our project.

As time passed, however, some of the alternatives being offered by the new archaeologists began to reach even some of us working in Greece. In terms of issues of interest to us, for example, Lewis Binford's "Post-Pleistocene Adaptations" (1968) offered a provocative new model for investigating the "Neolithic Revolution." But perhaps even more directly influential was the work of Colin Renfrew, who probably provided the most effective bridge in the practice and thought of prehistoric archaeology between Europe and North America. Indeed, his *Emergence of Civilisation* (Renfrew 1972) essentially brought the new archaeology to Greece. More than any other single work, the appearance and impact of that book represent a watershed in the practice of Aegean archaeology.

Among principles espoused by the new archaeology, we found the "systems approach" and the "concept of process" (Gibbon 1984) most appealing. "Processual archaeology," by its acknowledgment of the interrelatedness of the components of the archaeological record and the awareness that change in that record could be generated from within the system, seemed a reasonable extension of the ideas advanced by Grahame Clark years earlier. And the same may be said of the attention drawn to the relationship between human culture and the natural environment by the emerging (sub)discipline of cultural ecology or ecological anthropology. The spirit of "archaeology as human ecology," as Karl Butzer (1982) titled his book, has played an important role in the thinking of our project and the Argolid Exploration Project from very early on.

HISTORICAL TOPONYMY OF THE SITE

Despite the scattered (and largely unstratified) evidence of human activity during the historical periods revealed by our excavations[3] and the likelihood that Franchthi Cave lay near and presumably within the jurisdiction of the ancient town of Mases[4], its ancient name cannot be attested with any certainty. The most likely candidate, in my view, would be the headland (Greek *akra*) of Strouthous mentioned by the second century traveler Pausanias in his account (2.36.3) of the area.[5] That Roman Mases lay in the vicinity of modern Kiladha is now widely accepted (Jameson et al. 1994), and Pausanias clearly associates Strouthous with Mases.[6] This has led some modern observers, quite plausibly, to equate "Akra Strouthous" with the limestone headland now known as Franchthi.[7] Yet Pausanias's account is vague at this point (as often elsewhere), and Jameson (1953) is probably correct in asserting that he did not travel extensively in the area, and his first-hand knowledge of it was very limited. Jameson therefore is inclined to associate Cape Strouthous with the modern Cape Iria (Kavo-Iri).[8] In view of these uncertainties, the ancient name of the headland (and cave) of present-day Franchthi must remain an open issue.

It is not until much later, during the second Venetian occupation of the Morea (Peloponnesos) in the late seventeenth and early eighteenth centuries, that we encounter the first clear reference to the headland, but not yet to the cave itself. In a sketch map belonging to the cadastral survey of ca. 1705 (Topping 1976; Jameson et al. 1994), land parcelment and the location of several sites in the neighborhood of Kiladha Bay ("Porto Chiladia") are indicated, including "M[onte] Francti." This, then, is the later Venetian version of the present toponym, itself probably having older roots. In fact, it seems that "Franchthi" is an Albanized form of a fundamentally Greek word that reflects the Albanian immigration and influence in this part of Greece since the fourteenth and fifteenth centuries.[9]

After the Venetian period, there are no known citations of the place name until well into this century. It is curious that, while a few of the nineteenth-century travelers in the Peloponnesos appear to have been aware of the headland, none seems to have visited it, and no one refers to it by name. Even Miliarakis (1886), who is the first to mention the cave itself, describes its location in terms of Kiladha and the nearby islet of Koronis. He may not have visited the area. Yet his reference to the cave as a source of niter (used in making gunpowder) is another reflection of the extent of the mining activities to which the site has been subjected in recent times and helps to explain the considerable surface disturbance that we encountered in our excavations (see Chapter 2).

I know of only three published versions of the toponym prior to the commencement of our project: "Fragthi" (Army Map Service 1954), Φράγχθη (Royal Hellenic Navy 1958),[10] and "Frankhthi" (United States Office of Geography 1960). While the latter transliteration might now seem slightly preferable to our choice, ours has been around for some three decades and has gained wide currency in the archaeological literature. It would therefore be unwise to alter it at this point in time.[11] Yet, since our initial reports, all manner of variations have appeared in published references to the site by others. Perhaps the most amusing of them is that which appeared on a recent tourist map of the Peloponnesos published in Greece (Efstathiadi Group, 1970). The primary spelling is the transliteration preferred by us (i.e., "Franchthi"), while the Greek spelling (Φρανχθή; note also the stress accent is on the last syllable) is unique and seems to be a re-transliteration into Greek from our version.[12]

Until the time of our excavations, the site seems to have been known locally as the "Cave of the Cyclops" (Jacobsen 1969a) or merely the "Cave of Kiladha" (Andonakatou 1973). There is a cave or sinkhole on the small island of Koronis opposite Kiladha and the Franchthi headland. It also seems to have been known as the "Cave of the Cyclops" (Spiliopoulou 1965).[13] By the time I was first allowed to visit the island in June, 1976, the larger part of the sinkhole had been paved and transformed into a movie theater by the owners of the island, the Livanos family. It was therefore impossible to evaluate either its speleological or archaeological significance.[14]

In summary, then, very little was known about Franchthi Cave before we began our excavations in 1967. Ours were therefore the first formal investigation of the site, which is rather surprising given its size and striking appearance (especially when viewed from across the bay in the village of Kiladha). Perhaps this was due to the relative lack of interest in early Greek prehistory among archaeologists working in Greece as well as in cave archaeology in general prior to the commencement of our excavations. That situation has certainly changed in more recent times (witness, for example, the formation of a "Cave Ephoreia" in the structure of the Greek Archaeological Service).

NOTES

1. I had the good fortune to have been a graduate student at the American School of Classical Studies in Athens in 1963 while Weinberg was a resident faculty member preparing this study. Accordingly, we had numerous informal discussions about issues related to Aegean prehistory (especially the Neolithic period in Greece). Weinberg's 1965 fascicle eventually appeared virtually unaltered in the hardcover edition of the *Cambridge Ancient History*, Volume I, in 1970.

2. As a student of McDonald at the University of Minnesota in the late 1950s, I developed an early and long-lasting friendship with him. Thereafter, we communicated regularly and often and exchanged much in the way of information and ideas about what we considered the "proper" practice of archaeology in Greece. This cross-fertilization continued after his excavations at Nichoria and ours at Franchthi were under way, and members of both staffs interacted routinely.

3. E.g., Jacobsen 1969b, 1973, 1979. There are good indications that the cave served as a cult place from late archaic times into the Roman period. It is not yet clear, however, what cult(s) worshipped there, or if they were related to the ancient cult activity attested at the "temple terrace" on the southern slope of Franchthi headland. For a discussion of the latter (site #C 17), see Jameson et al. 1994:469.

4. On ancient Mases, see Meyer 1930. While significant enough to have been included in the Homeric Catalogue of Ships (Iliad 2.562) and still flourishing in the second century A.D., it is not clear that Mases ever attained the status of a polis (city-state) (Jameson et al. 1994:375–377).

5. Strouthous may be translated from ancient Greek as "sparrow" (or, perhaps, simply "bird"). It is noteworthy that, for anyone who has spent time in Franchthi Cave, one cannot help but be struck by its role as a haven for birds. Watching them swiftly winging their way in and out of the difficult crevices in the roof of the forepart of the cavern (the occasional younger or inexperienced one miscalculating and being dashed on the rocks) is a memorable sight indeed.

The earliest reference to Strouthous (and the only other known to me) is found in a Hellenistic (early second century B.C.) inscription recording a border dispute between Hermion(e) and Epidauros. The topographical problems in the inscription have been discussed (with full references) by Jameson (1953).

6. Pausanias traveled from Hermione on the east coast of the southern Argolid peninsula to Mases on the west coast. "From Mases there is a road on the right [i.e., to the north] to a headland called Strouthous . . ." (translation by W. H. S. Jones in the Loeb Classical Library edition of Pausanias, volume 1, p. 447). This certainly fits with our impression of the physical relationship between the Franchthi headland, on the north side of Kiladha Bay, and Mases (site #C 13: Jameson et al. 1994) on the south shore of the bay. But the subsequent distances given to the neighboring locations of Philanorion (probably modern Fourni) and Didymoi (modern Dhidhima) do not make sense.

7. E.g., Bursian (1872:92). Leake (1830:463) seems to be implying the same thing, though he mistakenly associates it with modern Koraka (Korakia). Later (1846:288–289) he changes his mind and locates Strouthous in the area of the Bay of Vourlia, some distance to the north of Kiladha, as do several other modern commentators on Pausanias. Cf. also Pouqueville 1827: 261.

8. Jameson 1953, and Jameson et al. 1994. It is interesting that Jameson (1953:166), clearly describing the headland of Franchthi ("a long ridge which descends steeply to the sea"), still does not refer to it by name. Nor does he mention the cave. He goes on, "The road mentioned by Pausanias as leading from Mases (in the Koilada plain) to Strouthous must have passed around this ridge to the east and continued around the projecting point of Akrotiri Salanti [north of Franchthi], as does a path today."

9. I owe this information, for which I am extremely grateful, to Professor Eric P. Hamp of the University of Chicago (personal communication 1973). Very simply, the word seems to be an Albanian diminutive (hypocoristic) of the Greek Φραγκος (="[W.] European"). (It is not the Albanian word for "cave," as Haller von Hallerstein [1986:188] writes.) I can offer no explanation as to why the headland should have been so designated.

On the period of Frankish influence in the southern Argolid and the Albanian settlement of the area, see Jameson et al. (1994). The latter (p. 121, n. 51) refer to Franchthi as "the Frankish Place," which does not seem to capture the full spirit of Hamp's etymology.

10. This continues to be the most common spelling in Greek and has served as the basis for most subsequent transliterations, including our own.

11. Hamp (personal communication 1974) recommended "Francthi" as an easier spelling to deal with, and this is occasionally seen in publications referring to the site.

12. F. W. Householder of Indiana University confirmed (personal communication 1986) that this is an un-Greek orthography, though perhaps appropriate for a foreign word.

13. I am grateful to J. M. Wickens for this reference.

14. Adonis Kyrou, a local antiquarian and valuable informant in the early years of the AEP, claims to have been the first to recognize the archaeological significance of our site (in 1959). He reports that in 1966 the new owner of Koronis Island, in the course of his renovation of the place, did a good bit of damage to preexisting remains on the island as well as to the area of the cave (Kyrou 1990:40, n.1). We know too that Kyrou himself used the cave as a party area for friends on more than one occasion prior to our arrival. All of this activity further explains the disturbed stratigraphy at the site (see above and Chapter 2).

CHAPTER TWO

Excavation History and Methodology

HISTORY OF THE EXCAVATIONS (T. W. J.)

The complex relationship between the Franchthi excavations and its "umbrella" program of fieldwork in the southern Argolid, the "Argolid Exploration Project" or AEP, has been outlined in the previous chapter. The purpose here is merely to summarize the history of those excavations and the attendant study seasons leading up to this publication series.

The excavations at Franchthi Cave began in late June 1967 under rather unusual circumstances. The site had been discovered during the previous summer, but, while something of its potential importance had immediately been recognized, we had no intention of beginning work there so soon. Those plans changed quite abruptly during the academic year 1966–67. When problems with the expropriation of land for excavation at Halieis became apparent (see Chap. 1), the AEP codirectors decided to have a trial season at Franchthi, given the unexpected existence of some funding and authorization from the Greek government. This trial excavation was conducted with a very small staff and with no expectation that further work would be authorized at the site. The initial season was therefore viewed as something of a temporary, stop-gap operation until the Halieis problems were resolved. The logistics of this decision were complicated somewhat by arrangements for housing. Accommodations had already been established in Porto Kheli, some 12 km by road from the site, and the situation was further complicated by the fact that the field director was also nominally responsible for the small excavation conducted by F. R. Matson at Lorenzo Kiln, some six km beyond Porto Kheli.

The logistical problems of the first season (June 21 to August 26, 1967) persisted into the second season, 1968. While it had been our intention to return to Halieis, the expropriation proceedings had again been delayed. Permission was granted to conduct a small trial excavation outside the Halieis city walls, but that lasted for only a short time. In view of the importance of the first season's findings, we sought permission to continue work at Franchthi, where we worked from June 22 through August 14. The staff (now somewhat larger) was again housed in accommodations at Porto Kheli.

The organizational complications of the first two seasons were resolved in 1969 with the restructuring of the project that brought Rudolph to Halieis and permitted Jacobsen to concentrate on the Franchthi excavations (see Chap. 1). The third season at Franchthi lasted for two months (May 19 to July 19) and was greatly facilitated by the transfer of staff housing and work space from Porto Kheli to Kiladha. Excavation at the site was confined to the cave itself during the first three seasons, 1967 to 1969.

The restructuring of the project in combination with overall funding limitations led to the creation of a staggered excavation schedule for the Halieis (both land and underwater) and Franchthi teams in most of the following years. In 1970 fieldwork was restricted to Halieis, and the Franchthi staff had a study season at the Archaeological Museum in Navplion. In 1971 the Franchthi excavations were resumed (fourth season), and less work was done at Halieis. The fourth season at Franchthi lasted from June 14 to August 7 and included the initiation of work outside the cave, in the area that came to be known as "Paralia." The excavations were at Halieis in 1972, and the Franchthi team held another study season. Some members of the staff took part in an intensive surface survey of the area around the cave under the direction of Jameson, Jacobsen, and Dengate (see Chapter 1).

The fifth and sixth campaigns at Franchthi were held in 1973 and 1974, each lasting more than two months (June 6 to August 12 and April 8 to June 28, respectively). These were not only our longest seasons, but our largest (in number of personnel) as well. Considerable attention was given to the enlargement of the exposures on Paralia during these seasons. It should be noted that the spring campaign in 1974, as indeed portions of certain other seasons, permitted the sharing of personnel between the Halieis and Franchthi teams.

The 1975 season was again devoted to the study of the finds in the Navplion Museum, where limitations of space necessitated that the large group be divided into two smaller components, each working one month (between May 15 and July 15). That season was also used to develop plans for our proposed seventh and last major campaign in 1976.

The 1976 season was devoted to a variety of undertakings, including additional excavation on Paralia and in the cave, detailed balloon photography, further analysis of previous seasons' finds, and various geological studies, including extensive sampling of the cave sediments by W. R. Farrand. This campaign lasted for exactly two months, from May 31 to July 31.

The bulk of our time and effort since the last field season has been devoted to the identification, classification, and analysis of the excavated remains. This has been a formidable undertaking from the point of view of those responsible for the study of very large bodies of material (e.g., the zoological, botanical, ceramic, and lithic samples) and, as always, could be accomplished only within the constraints of available funding and time free from other responsibilities. Moreover, constraints were imposed by the limited numbers of hours that the Navplion museum was open to us—only from 7:00 A.M. to 2:00 P.M., Monday through Friday, on a regular basis. While the pace of these studies has been necessarily deliberate, progress toward publication has been steady. Our progress was punctuated by three significant events: 1) a symposium of core personnel on the project, held in Bloomington, Indiana, in the autumn of 1978, 2) a final ("clean-up") season in the field during the summer of 1979, and 3) a second symposium in Bloomington in October 1982.

The principal results of the 1978 symposium may be summarized as follows. The majority of the publication assignments were confirmed. A working (tripartite) format for the final publication was agreed upon (see Jacobsen and Farrand 1987: 9–10), and a contract was signed with a publisher. Considerable discussion was devoted to planning for the 1979 season and, especially, the attendant implications for an overall "phasing" of the site (see Jacobsen and Farrand 1987:8–9).

The primary objective of the 1979 field season, then, was to establish the stratigraphic basis for the site phasing. To that end, most of the season was devoted to a reexamination of the sediments and stratigraphy of the cave and Paralia and the preparation of final section (profile) drawings with conventions appropriate to the two areas. A second major undertaking during that season was the initiation of investigations concerned with the identification of past shorelines and determining the physical limits of the site. This was undertaken by a geophysical survey of the offshore area under the general direction of Tj. van Andel (van Andel et al. 1980).

The summers of 1980 and 1981 were again taken up in large part by studies for final publication, but a brief program of coring was conducted by J. A. Gifford in the narrows between the cave and Koronis Island in August, 1981. This served as a follow-up to van Andel's work in 1979 and was designed to test our hypothesis (proposed in the early years of excavation of Paralia) that the Neolithic settlement outside the cave had been partially submerged by postglacial sea-level rise.

In October 1982, a second gathering of most of the people involved in the project took place in Bloomington. Those meetings were again a notable success and stimulated, perhaps even more than did the 1978 symposium, discussions of a truly interdisciplinary character. The deliberations focused on two broad topics: 1) the nature of the final publication, where certain refinements in the tripartite format were agreed upon, but no dramatic changes were introduced, and 2) the overall phasing of the site. As regards the latter, the basic philosophy of the 1978 symposium was reaffirmed, but the collective experience of the various contributors, who had in the meantime been working with the stratigraphic sections in combination with sequences of excavated units, revealed problems of detail. While those problems required certain compromises in the establishment of sequences for the various bodies of material from the two major excavated areas (cave and Paralia), the group remained firmly committed to an initial approach of independent phasing of different categories of remains, as illustrated hypothetically by Jacobsen and Farrand (1987:Figure 5) and in a preliminary form in Farrand (1993:Figure 6). It became clear to us that the achievement of an integrated phasing of the site as a whole could probably not be attained before the Level Two (synthesis) publication.

The only field work conducted since the 1982 symposium was a more extensive program of coring in Kiladha Bay in May and June 1985. This project, planned as an immediate follow-up to the coring in 1981, had been delayed because of problems of authorization by the Greek government. The work was again undertaken by Gifford and a small team of trained scuba divers (Gifford 1990).

TRENCH LOCATIONS

The excavations inside Franchthi Cave were conducted in eight seasons from 1967 through 1976, as summarized in the previous section. Excavation proceeded simultaneously in several trenches labeled A, F, G, and H with various subdivisions and combinations, as described in this chapter (Figure 2.1). The trenches are roughly parallel to the long axis of the cave, which is N37°W, and to each other, but they are individually oriented to accommodate the topography of the cave floor. The long axis of the Trench A and F composite is N49°W, the H-H1-H2 composite is N57°39'W, and Trench G is parallel to the west wall at N32°W. By convention and for simplicity, excavators referred to the end of each trench nearest the cave opening as "north," a convention that is followed in this fascicle.

These trenches are situated in the easily accessible, relatively flat area of the cave that extends inward from the present drip line to the edge of a huge, chaotic pile of limestone blocks, many measuring several meters across, that collapsed into the cave in post-Neolithic times (Figure 2.1; Plates 3b, 4a). The collapse opened a very large "window" to the sky in the middle of the cave (Plate 1b), which presumably was not widely open, if at all, prior to the end of the Neolithic occupation of the cave. The presently accessible cave floor is about 35 m wide by 38 m long and has an area of about 1350 m². The surface dimensions of the four trench complexes total about 136.5 m², or about 10% of the total accessible area.

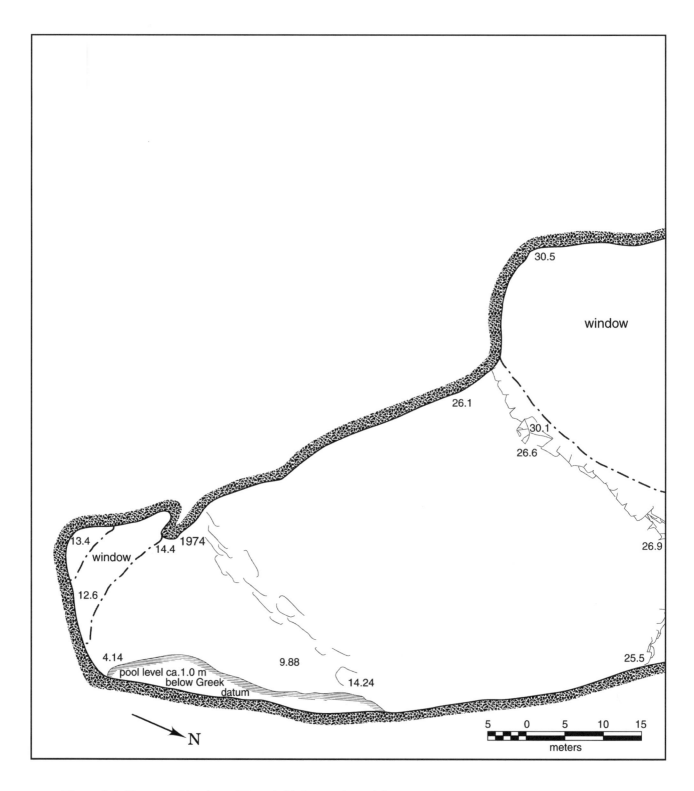

30.5

window

26.1

30.1

26.6

13.4

window

14.4   1974

12.6

26.9

4.14

pool level ca.1.0 m
below Greek
datum

9.88

14.24

26.9

25.5

N

5 0 5 10 15

meters

Figure 2.1. Topographic plan of Franchthi Cave, adapted from Jacobsen and Farrand (1987:Plate 2).
Dashed contour lines in the cave area are contours on the bedrock surface above the cave.

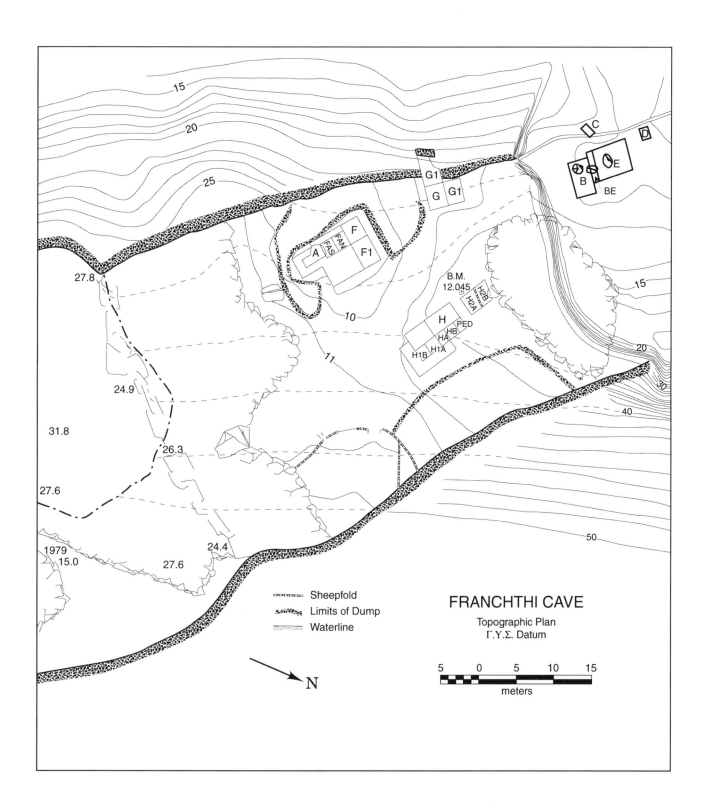

15

20

25

G1

G1

G

G1

F

FAN
FAS

A

F1

27.8

B.M.
12.045

H2B
H2A

H

HB
PED
HA

H1B
H1A

15

24.9

31.8

26.3

27.6

11

10

20

30

40

1979
15.0

24.4

27.6

C

D

B

A

E

BE

50

Sheepfold

Limits of Dump

Waterline

FRANCHTHI CAVE

Topographic Plan
Γ.Υ.Σ. Datum

N

5   0   5   10   15

meters

Trenches B through E were shallow excavations located 4 to 15 m outside the present cave entrance (Figure 2.1). They were excavated in the preliminary season of 1967, and most of the finds were examined in the field and not curated. In 1969 the balk ("BE") between Trenches B and E was excavated more carefully. These exploratory trenches were 1 m deep at the most, terminating on a cemented mass of large rock debris, presumably resulting from collapse of the cave brow in pre-Neolithic time. The looser material above the mass of rocks contained numerous marine shells, some lithic artifacts, and potsherds of ceramic phase FCP 1 (Vitelli 1993:33).

Upon close inspection, one may note that the dimensions and shapes of certain trenches as shown in this fascicle differ somewhat from those in Fascicle 1 (Jacobsen and Farrand 1987). Minor discrepancies have been corrected, and the drawings herein should be considered to be the more definitive.

EXCAVATION SEQUENCE

Trenches A, F, and G were begun in 1967. As explained in Chapter 1 and above, the 1967 season was intended as a simple exploration of the cave without any firm intention of continuing work there. Trench A was completed in that season, and Trenches F and G were terminated in 1968. Plans of these trenches are shown in Figures 2.2 and 2.4. (see also Plate 3a), and vertical sections showing the chronological sequence of excavation are in Figures 2.3 and 2.5. Trench A was begun over a large 3 by 5 m area, but a very large block of limestone roof fall (called the "great boulder") was encountered early in the excavation and was an impediment to systematic excavation until it was broken and removed at a depth of about −2.5 m. At that time the area of Trench A was reduced to approximately half the initial size (Jacobsen 1969a) and took on an irregular shape because of the disturbance associated with removal of the limestone block (Figure 2.2). In fact, the north end of Trench A overlapped the area that was later to become Trench FAS, down to a depth of −5.88 m, as indicated in Figure 2.3.

All depths given in this chapter are depths below the established datum for a particular trench, and the datum was essentially at the level of the cave floor in the vicinity of each trench. These datums, in meters above sea level, are 9.99 for Trench A, 9.97 for FA, 10.09 for Trench A, 11.32 m for Trench GG1, 12.29 for HH1, and 12.045 for H2, which is also the official cave datum shown on Plate 2 of Jacobsen and Farrand (1987) and Figure 2.1 herein. Additional information on the topographic survey of the cave by F. Cooper is found in Jacobsen and Farrand (1987:11–14).

Trench F began as a 3 by 3 m area, but in 1968 it was expanded by the addition of section F1, 3 by 3.5 m in area (Figure 2.2). However, much loose rock and what turned out to be modern disturbance was encountered in Trench FF1. Thus, below a depth of −3.6 m, excavation was limited to a much smaller area, about 2.5 by 1.75 m, in the SW corner of FF1. That smaller area was further subdivided into four quadrants, FF1:A, B, C, and D, which were excavated down to about −4.8 m.

Trench G, just inside the drip line of the cave, had a more complex history (Figures 2.4 and 2.5). The initial Trench G was 3 by 3 m in area, and it was excavated down to about −5.7 m. Near the end of the 1967 season, the trench was widened, as "G Extension," one meter to the south and all the way to the cave wall, 1 to 2.5 m, on the west, but was dug only to a shallow depth of about −1.2 m in 1967. In 1968, section G1, 2 by 3 m, was added on the north side of the initial trench and was then expanded to include the area of G Extension. This area was sometimes referred to as GG1, although it did not include the area of the original Trench G

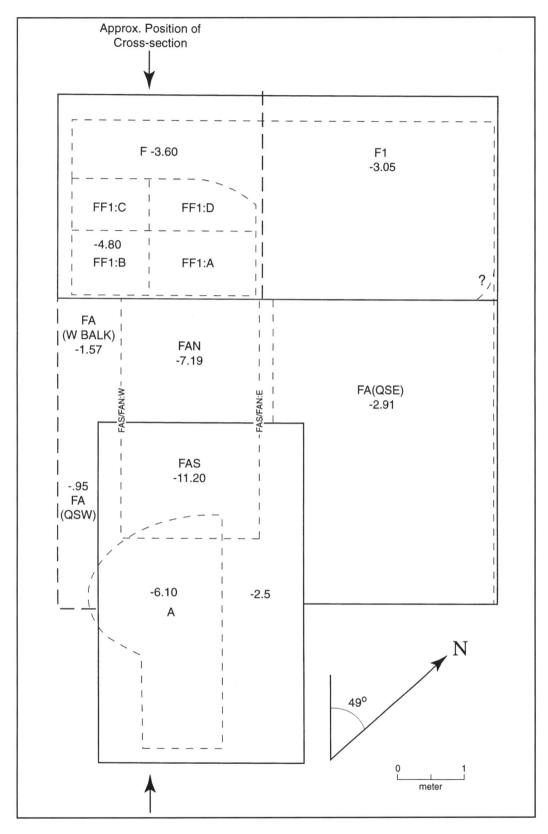

Figure 2.2. Plan of the subtrenches in the A-FA-FF1 complex. See text for explanation of symbols. The depth of the bottom of each subtrench is indicated in meters below the trench datum. The approximate position of the vertical cross-section shown in Figure 2.3 is indicated by the bold arrows.

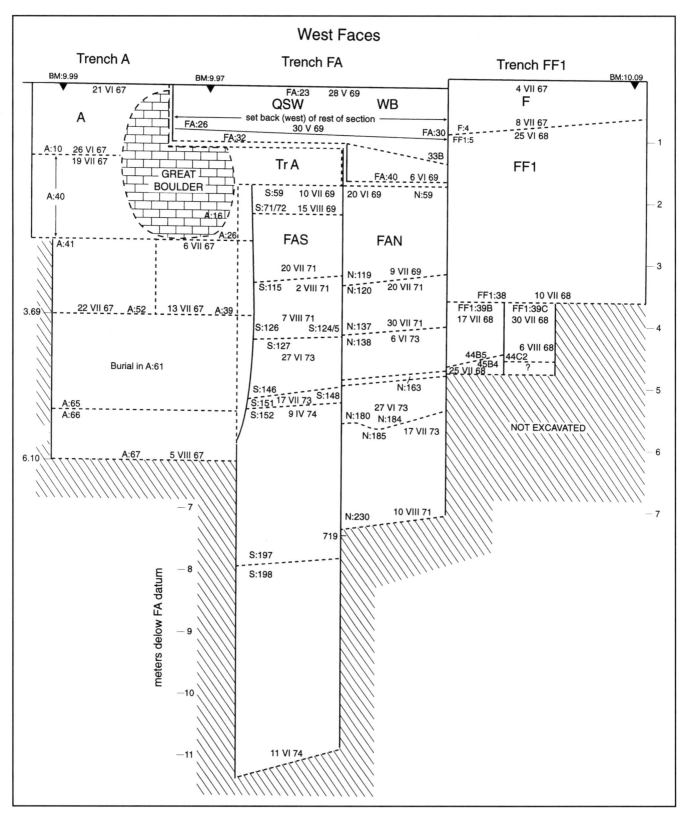

Figure 2.3. Schematic vertical section across the Trench A-FA-FF1 complex showing selected unit numbers and the dates when excavation of that section of the subtrench was begun and ended.

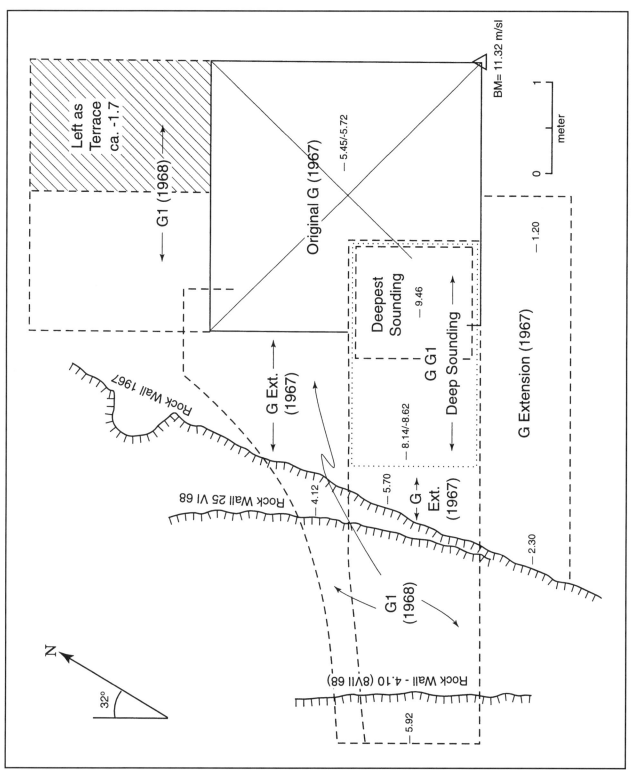

Figure 2.4. Plan of the Trench GG1 complex. The depth of the bottom of each subtrench is indicated in meters below the trench datum. The position of the cave wall is shown for three different depths indicating the westward inclination of the cave wall with depth.

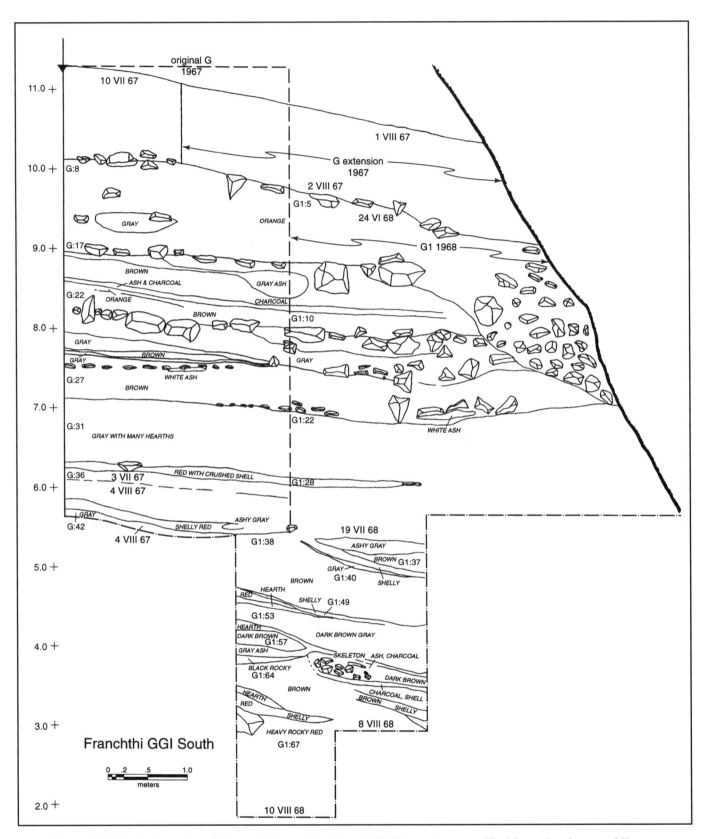

Figure 2.5. Vertical section of the south face of the Trench GG1 complex, modified from Jacobsen and Farrand (1987, Plate 21). Conventions as in Figure 2.3.

except in the deep sounding below −5.7 m (Figure 2.5). During excavation, Trench G1 expanded westward against the cave wall, because the cave wall widened outward with depth. When G1-G Extension reached a depth of about −5.7 m, which was the original bottom of Trench G, its area was narrowed to 1.5 m wide. This narrow, "deep sounding" was excavated rapidly in the hope of reaching sea level by the end of the 1968 season. The deepest point, in the southwest corner of the original Trench G, was −9.46 m, still about 2 m above sea level.

A broad area between Trenches A and F, referred to as "FA balk," "F/A," or simply "FA," was excavated piecemeal beginning in 1969, as follows (see Figure 2.2). Initially relatively broad areas, Quadrant Southeast (FA-QSE) and Quadrant Southwest (FA-QSW) were excavated down to approximately −1.6 and −2.9 m, respectively, and ultimately they were joined through West Balk (FA-W balk) and East Balk (FA-E balk). Finally, deep excavation over a more limited area (roughly 2 by 3.5 m) in two contiguous trenches, FA South ("FAS") and FA North ("FAN") was begun late in the 1969 season and carried down to −7.19 m in FAN in 1973 and to a point where groundwater (sea level) was encountered, or a depth of −11.2 m, in FAS in 1976 (Figure 2.3). Excavation unit numbers were assigned serially throughout all FA subtrenches.

The Trench H-H1-H2 composite was begun in July 1968 in a 4 by 4 m area called Trench H (Figures 2.6 and 2.7), the north edge of which is about 15 m inside the drip line of the cave. The drip line in this area is coincident with a prominent pile of limestone blocks that collapsed from the brow of the cave. At a depth of about −3.5 m, Trench H was reduced to an area of 1.87 by 2.5 m in the north half of the trench because of "treacherous rocks" protruding from the walls and the presence of a large mass of travertine-cemented rocks in the south half of Trench H. At a depth of about −5 m, this area was divided in two adjacent subtrenches, HA and HB, each 1.8 by 1.25 m. Trench HA was excavated down to about −7.6 in 1968, and HB, begun in 1968, was terminated at −6.65 m in 1969. Also in 1969 and 1971, an area in the northeast corner of Trench H, called H Pedestal (or "H Ped"), was excavated down to about −4 m.

Early in the 1969 season, excavation of the area adjacent to the south side of Trench H was begun. This was called Trench H1, which was initially 3 by 4 m in area, and excavation over the entire 12 m² was carried down to about −3.5 m. At this point, Trench H1 was divided into four quadrants of unequal size, labeled H1A, H1B, H1C, and H1D, clockwise beginning with the northeast corner. Quadrants C and D were not excavated. Quadrants H1A and H1B were laid out initially as areas of 1.5 by ca. 1.65 m, but H1A was expanded northward to include a narrow 0.30 to 0.35 m area in the southeast corner of Trench H that was left unexcavated when areas HA and HB were laid out (Figures 2.6 and 2.7). Quadrant H1A was excavated down to −9.26 m in 1971. Only the upper 0.5 m of Quadrant H1B was dug in 1969, but another 4 m was excavated in the 1973 season, and the quadrant was deepened again in 1974 and 1976, finally reaching a depth of −9.71 m. Given the complications of numerous, interlocking limestone blocks (roof fall) at the base of both Quadrants H1A and H1B, excavation of the lowest part of H1B expanded into the south part of H1A, which had been left undug in 1971, as shown in Figure 2.7.

In 1976 the area called H1 Terrace was cleared along the east and south sides of Trench H1 in order to provide better working conditions for the excavation of the deep portion of Quadrant H1B. Since it was already known that the upper levels of Trench H1 were badly disturbed, the excavation of H1 Terrace was carried out rapidly, and the sediment was dry sieved through a McBurney Shaker and not sent to the water-sieving operation (see below for additional information on sieving; also Diamant 1979). Also the excavation units were unusually thick, commonly 20 to 30 cm and even as much as 50 to 60 cm, compared to normal thicknesses

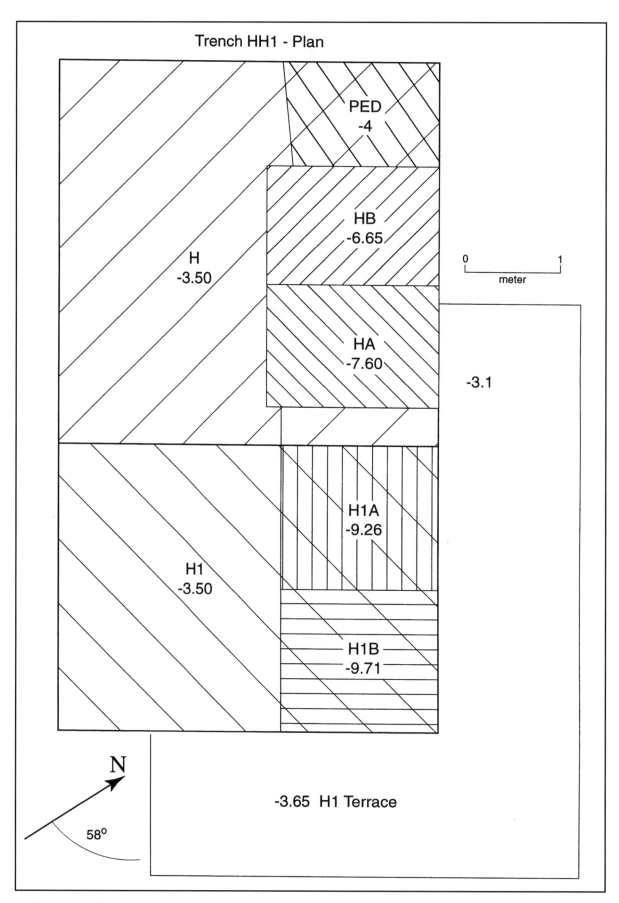

Figure 2.6. Plan of the subtrenches in the Trench HH1 complex. See text for explanation of symbols. The depth of the bottom of each subtrench is indicated in meters below the trench datum.

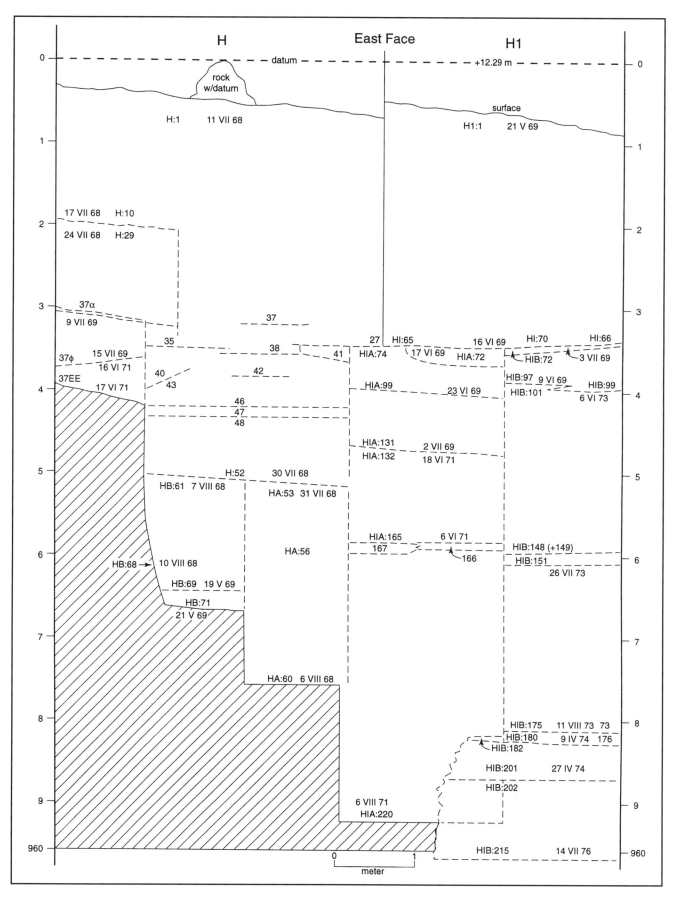

Figure 2.7. Schematic vertical section of the Trench HH1 complex. Conventions as in Figure 2.3.

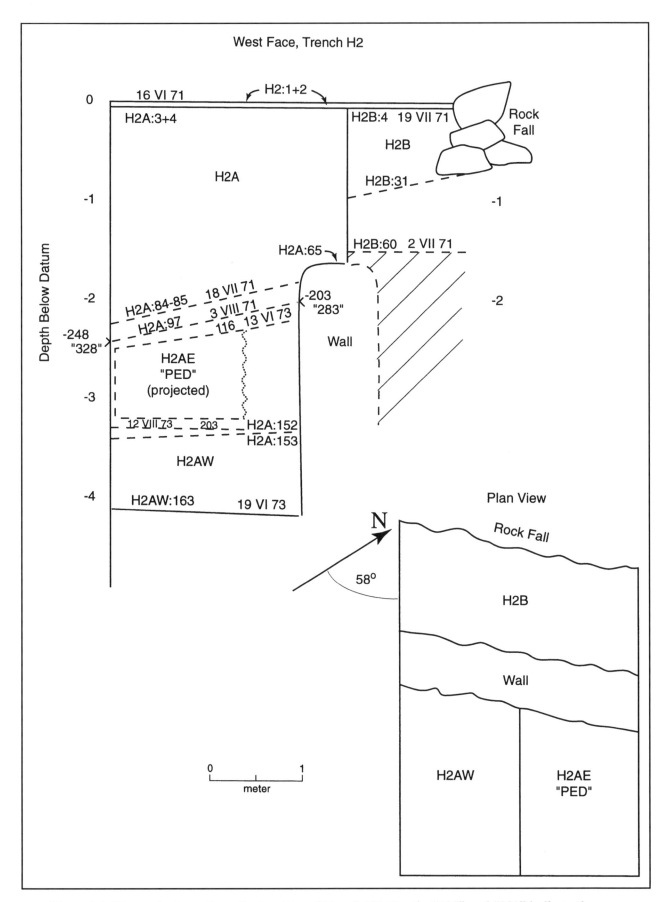

Figure 2.8. Plan and schematic vertical section of Trench H2. Depths "328" and "283" indicate the nominal but incorrect depths on the original section drawings. See text for further explanation. Other conventions as in Figures 2.2 and 2.3.

of 5 to 10 cm or less, to expedite the clearing of this area. The excavators believed that they were into "pure" (undisturbed) Middle Neolithic levels in the lowest excavation units on H1 Terrace; see also Vitelli (1993:Plan 2).

Trench H2 can be considered part of the composite of H trenches, although it is disjunct from the HH1 group. Trench H2 is in line with HH1, and its south edge is only 1.6 m from the north edge of Trench H. Its north edge abuts the rock fall under the brow of the cave. Trench H2 was begun in 1971 with an area 2.5 by approximately 3.5 m (Figure 2.8); its north edge was irregular because of the rock fall. At a depth of only a few centimeters, the trench was divided into two unequal parts, H2A in the south was 2.5 by 2.5 m, and H2B was the smaller part to the north against the rock pile. At a depth of about −1.6 m, a rock wall constructed by Neolithic inhabitants was encountered roughly along the line separating H2A and H2B. Beginning with Unit H2A:86 at a depth of −1.89 to −2.29 m, the north limit of H2A was specified to be the south face of this wall, which trends diagonally across the H2 trench (Figure 2.8; see also Figure 2.1). The area of Trench H2 north of the Neolithic wall was not excavated.

Early in the 1973 season, H2A was subdivided into two subequal halves. The west half, H2AW, was dug first, terminating at about −4.2 m depth. The east half, H2AE or H2A Ped, was excavated only to a depth of about −3.35 m. In Figure 2.8, H2AE is shown projected onto the H2AW section.

In compiling the section drawings for Trench H2, an error in field notes was discovered, as reviewed in Fascicle 1 (Jacobsen and Farrand 1987:25). An irrecoverable error was made in recording the depths of subdatums used in 1973, which implied a gap of about 0.80 m between the last unit excavated in 1971 and the first one in 1973. As explained in greater detail in Fascicle 1, we were able to piece together various bits of information to make a close estimate of the correct depths for the 1973 excavation units. An uncertainty of no more than 2 or 3 cm remains, and it is not considered a significant handicap in interpreting the sequence.

EXCAVATION METHODOLOGY

The excavation methods used at Franchthi Cave were essentially the same throughout the years, but improvements in quality of recording and plans, attention to detail, color notation, systematic depth recording, etc., occurred with the experience accumulated by the excavating staff. The basis for recording the recovered materials was the **excavation unit**, which is a three-dimensional body of sediment removed sequentially by the excavators within each trench or subdivision of a trench. (See also Jacobsen and Farrand, 1987:16, 25) The excavation unit, generally referred to simply as "unit," is roughly horizontal, but follows the natural stratification to the extent possible, and it includes all the physical substance removed from a specified portion of the trench, between two surfaces (which may be irregular in form). Successive units may be either vertically or laterally contiguous, and they may encompass the entire area of the trench or only a small, circumscribed area within the trench, such as a pit or hearth. (In the 1967 and 1968 seasons the units were called "baskets," according to traditional excavation terminology in Greece, and each basket usually comprised several passes across a given area.) A given excavation unit may have been dug in a single pass across the trench bottom, or in successive passes of 1- to 5-cm thickness if no obvious changes in sediment type appeared. Individual passes within a single excavation unit were given formal numbers only in Trenches FF1, HA, and HB (see next paragraph).

Units are numbered serially from the top down for each trench or subtrench, in the form of FAS:193 or H:53A2, that is, trench number-colon-unit number and (in some cases) pass

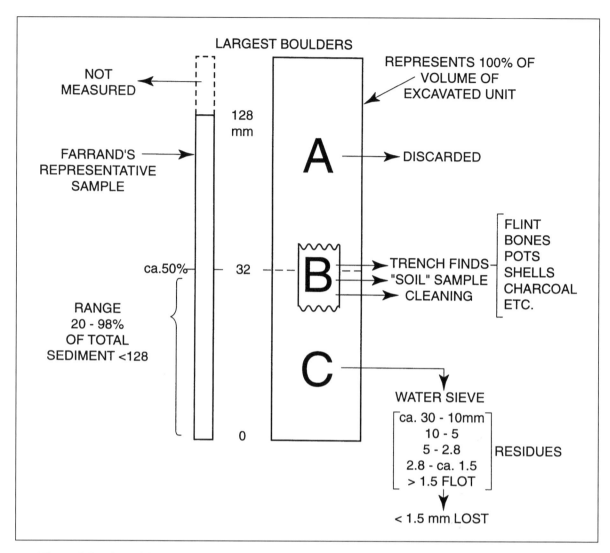

Figure 2.9. Disposition of the total sediment removed from a given excavation unit. The right side shows the fate of various components as controlled by the excavators. The size limits for the B component were somewhat variable. The left side shows the disposition of sediment removed for this study. Any material >128 mm (−7 phi) was noted, but not collected, for statistical reasons. The sediment fraction <32 mm (−5 phi) averaged about 50%, but it was quite variable, ranging from 20 to 98% of the collected sample.

number with no spaces included. A complete list of unit numbers is given in Jacobsen and Farrand (1987:21). Key unit numbers for the top, bottom, and interseason gaps in each trench are indicated on the schematic sections (Figures 2.3, 2.5, 2.7, and 2.8) included in this chapter. In general, a new number was given each time the excavators recognized any kind of change in the character of the sediment being exposed, e.g., color, grain size, consistency, wetness, artifact, bone, or shell density. In some trenches, especially HA, HB, and FF1 Quadrants A, B, C, and D, successive passes in a given unit were labeled 1, 2, 3, etc., such as in H:60A10, where the A indicates Quadrant HA and 10 is the tenth pass in unit 60. In H Ped only one unit, 37, was designated, and it was subdivided into passes labeled with uppercase Greek letters A, B, Γ, Δ, etc., e.g., H:37Σ.

It is important to know the shape and extent of the excavation unit in plan view and its vertical dimensions relative to a datum. These dimensions were not carefully recorded, if at all, in the first year of the Franchthi excavations, but considerable progress was made in subsequent seasons. Ideally a large number of depth measurements should be recorded if the surface of the unit is not planar, but in many cases only the depths to each corner of the trench were recorded in the early years. Moreover, if the excavation unit does not extend to the corner(s) of the trench, it is not always clear from the notebooks exactly where the measurement was taken, unless a careful plan was drawn of the surface of the unit. Carefully drawn plans became common in 1971 and later. In many cases, "sweep-o-grams"—roughly drawn plans of the trench bottom showing the area(s) included in the excavation unit—were included to show in a general way what part of the trench was included in a given unit.

Theoretically, excavation units should reflect the natural stratigraphy of the deposits; that is, their surfaces and lateral limits should be controlled by natural variability, such as sedimentary stratification, color differences, anthropogenic features (hearths, pits, floors, middens), and prehistoric or pre-excavation disturbance (rodent holes). In some cases, excavation units may be arbitrary. In the case of a natural layer that is considered too thick to remove in its entirety as a single excavation unit, arbitrary passes of 5 or 10 cm may be removed, preferably parallel to whatever natural surfaces can be detected. In other cases, no natural stratigraphy may be evident because of very uniform sedimentation, various forms of turbation, or disturbance. Thus, relatively thin, horizontal units should be removed. An excavation unit should be no thicker than the maximum amount of sediment the excavator is comfortable in removing at one time in order to preserve a detailed stratigraphy. It is important to recall that the rate of sedimentation in many caves is rather slow. In Franchthi Cave the rate varied from 1 to 2 cm per 100 years to as much as 250 cm per 100 years (Farrand 1988:315, and Chap. 6 below). At the slow end, an excavation unit of 5 cm thickness could encompass 250 to 500 years accumulation. At times of very rapid sedimentation, 5 cm may include only 2 years accumulation. These rates are, of course, averages.

The excavation unit is the smallest or most detailed provenience for all the archaeological materials recovered by sieving operations. Sieve residues cannot be repositioned into the stratigraphic matrix with any greater precision than that of its entire excavation unit. Only piece-plotted objects can have a more precise provenience, but very few objects were piece-plotted at Franchthi. Some larger items are shown on plans—large potsherds, figurines, large bones (including burials), charcoal, manuports, etc, but this procedure was not systematically employed, especially in the earlier years of the excavations.

VOLUME OF SEDIMENT EXCAVATED AND ITS DISPOSITION

Nearly 550 m³ of sediment was excavated from inside Franchthi Cave in eight seasons. That includes all the fine sediment, small and large rocks (up to a meter or so diameter), and all the archaeological objects (lithics, ceramics, animal bones, plant remains, human bones, manuports, etc.) The volumes removed from each trench and subtrench are given in the table below.

Table 2.1. Volume of Sediment Excavated in m³

TRENCH	NEOL	MESO	PALAEO	TOTAL
GG1	99.51	34.17	2.57	136.25
FF1	63.90	1.40	0	65.30
A	50.60	3.40	0	54.00
FA	66.88	17.33	11.49	95.70
HH1	96.32	18.50	21.38	136.20
H1 Terrace	29.43	0	0	29.43
H2	23.24	1.76	0	25.00
TOTALS	429.88	76.56	35.44	541.88

SUBTRENCH	NEOL	MESO	PALAEO	TOTAL
G & G EXT	52.98	8.37	0	61.35
G1	46.53	25.80	2.57	74.90
FF1	63.90	1.40	0	65.30
A	50.60	3.40	0	54.00
FA:QSW	2.59	0	0	2.59
FA:QSE	25.23	0	0	25.23
FA:WB	6.28	0	0	6.28
FA:EB+FA Balk	12.72	0	0	12.72
FAN	9.35	8.40	0	17.75
FAS	10.71	8.93	11.49	31.13
H	57.11	4.11	0	61.22
HA	0	2.21	3.24	5.45

Table 2.1 (Cont.)

SUBTRENCH	NEOL	MESO	PALAEO	TOTAL
HB	0	2.59	0.92	3.51
H Ped	1.71	0	0	1.71
H1	34.80	0	0	34.80
H1A	1.22	4.52	8.07	13.81
H1B	1.48	5.07	9.15	15.70
H2, H2A, H2B	23.24	1.76	0	25.00

The sediment excavated with each excavation unit was disposed of as shown in Figure 2.9. Of the totality of the sediment encountered, all the rocks and boulders larger than about 30 mm were discarded (Fraction A in Figure 2.9). (Thirty millimeters is a rough estimate of the lower limit for this material, because it was not rigorously specified.) Fraction B included all the archaeological finds removed during excavations, ranging from large potsherds and bones (animal and human) down to charcoal and lithic tools and fragments (chert and obsidian). The upper and lower size limits for material in fraction B were not standardized. Many of the microlithic artifacts were not detected in the trench and thus not removed by the excavators, given the coarse and commonly cohesive nature of the sedimentary matrix, and, for this reason, were not shown on plans (Perlès 1987:18). They were recovered in the sieves. Fraction B also included a sediment sample collected by the excavators, and any loose debris removed in cleaning the floor or walls ("scarps" in the excavators' parlance) of the trench. The latter debris was discarded, owing to lack of context. The remainder, fraction C, was sent to the sieving operations. In 1967 and 1968 throw screens and shaker sieves were used, but in 1969 water sieving was introduced and was utilized for most units during the remainder of the excavations. The sieving operations and the type of sieving for each excavation unit are detailed by Diamant (1979) and also reviewed in Perlès (1987: 17–18).

In summary, eight seasons of excavations in Franchthi Cave removed almost 550 cu m of sediment from three large trench complexes that sampled about 10% of the presently habitable cave floor. The sediment was removed in relatively thin "excavation units," delimited to the extent possible by natural stratigraphy. Larger objects were recovered from the trenches by the excavators. Smaller objects, especially most of the microlithic chipped-stone artifacts and all the microfaunal and palaeobotanical remains, were recovered by sieving, notably by water sieving in 1969 and later.

CHAPTER THREE

Geographical and Geological Setting

INTRODUCTION

The geographic, geologic, and climatic setting of Franchthi Cave is explained in detail in Fascicle 2 of this series (van Andel and Sutton 1987), which also includes a description of the soils and water resources of the area. A geologic map of the area surrounding the cave was prepared by C. J. Vitaliano and is published in Fascicle 1 (Jacobsen and Farrand 1987:Plate 1). Discussion of the geologic history by Vitaliano is found in Fascicle 2 (van Andel and Sutton 1987:12–17). This information will be repeated here only to the extent that it applies directly to the cave and to the interpretation of the cave sediments.

LOCATION AND CLIMATE

Franchthi Cave is located at the northwest end of a rocky headland across a small bay from the village of Kiladha (Plates 1a, 1b) in the southwestern corner of the Argolid Peninsula in the Peloponnese. The headland is a relatively small (about 1.7 km long NW-SE and 1.0 km across) mass of barren limestone (Figure 3.1), similar to most of the other hills of the southern Argolid. The summit of the headland is 176 m above sea level, and the headland is completely surrounded by low-lying, unconsolidated Quaternary deposits, mostly alluvium and slope deposits.

"The climate of the southern Argolid, one of the driest in Greece, is typically Mediterranean with its hot, dry summers and cool, moist winters" (van Andel and Sutton 1987:5). Local climate data are scarce, but records from Navplion, some 33 km to the northwest, indicate average temperatures of 10°C in January and 27°C in July. Average annual precipitation is about 500 mm, which falls mostly as winter rain. Occasional snow that falls on uplands above 300 m does not remain long on the ground. Strong winter winds blow mainly from the west and northwest directly into the mouth of Franchthi Cave. For more details see van Andel and Sutton (1987:5–8).

BEDROCK GEOLOGY

The Franchthi headland is underlain by Cenomanian (Lower Cretaceous) limestone, informally called the "Franchthi limestone" by Vitaliano. This limestone is mostly gray to black,

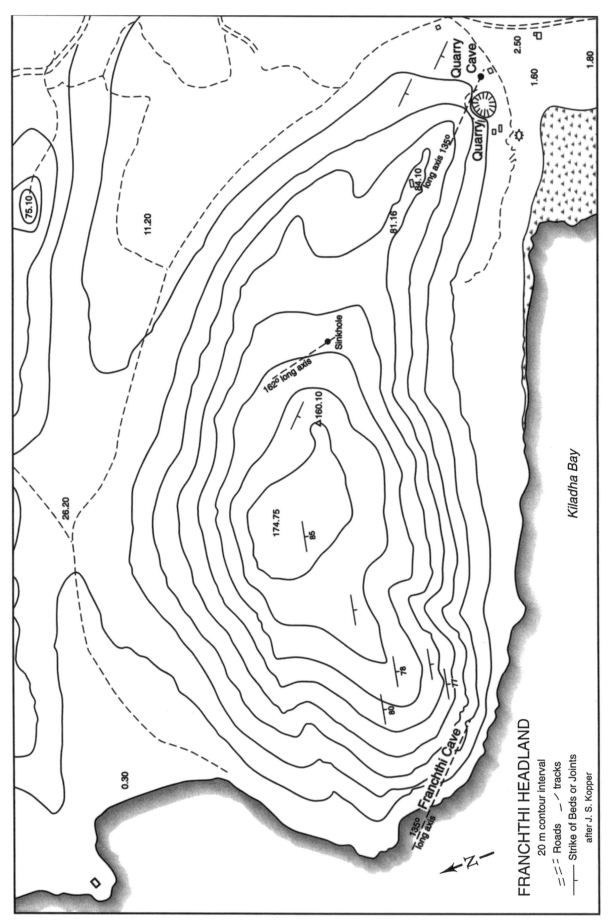

Figure 3.1. Topographic map of the Franchthi headland, adapted from an original by J. S. Kopper. Spot elevations are given in meters above sea level. The ESE-WSW strike and steep dips to the south are shown by symbols in the vicinity of the cave. Note also the parallelism of the long axes of Franchthi Cave and the quarry cave.

thick-bedded, and contains thin stringers of blue chert. It is characterized by Vitaliano as an "exceedingly fine-grained mass of calcium carbonate mud enclosing fragments of limestone and shells, and pellet-like aggregates of calcium carbonate" (van Andel and Sutton 1987:14). The limestone is highly fractured, and many of the fractures are filled with coarse-grained calcite. Conspicuously fractured zones are visible in some areas inside Franchthi Cave, for example, the far wall of the innermost end of the cave (Plate 6b). The individual rock fragments outlined by the fractures appear to be roughly the same size and shape as many limestone fragments that ultimately ended up in the sediments on the cave floor (Cf. Plate 6a).

A narrow belt of thin-bedded sandy limestone and calcareous marls of Paleocene to Eocene age occurs low on the southwest (bay-side) flank of the headland, overlying the Franchthi limestone. Although most of the uplands in the immediate area are underlain by limestone, important occurrences of an ophiolitic rock complex underlie the Fourni valley northeast of the Franchthi headland, and the terrain across the bay, just west of Kiladha village. This Jurassic ophiolite series comprises mafic rocks, probably originally diabase or peridotite, highly altered to serpentine, along with black and brown cherts, and purplish, thin-bedded limestones (Vitaliano in van Andel and Sutton 1987:13). Fragments and minerals of the ophiolite rocks occur in the cave sediments (see below), mostly introduced through human agency, although the finer grain sizes may have been blown in by the wind. The contact between the ophiolite series and the overlying Franchthi limestone may lie just below sea level along the Paralia. A number of springs occur in that area, and they are most likely related to the difference in permeability between the fine-grained serpentine and related rocks below and the fractured, somewhat porous limestone above. One of these springs (Figure 3.2) was tapped for a source of freshwater for the water-sieving operation (Diamant 1979:212–213), and presumably it was available to the prehistoric inhabitants of the area. Vitaliano had noted that fragments of the ophiolite rocks occur in the lower part of the Franchthi limestone, which suggests that the ophiolite series must lie closely beneath the limestone. This being the case, one can surmise that the ophiolite rocks occur under much of Kiladha Bay, below the young (Holocene) marine sediments (see also Gifford 1990).

KARSTIC PHENOMENA

In a region rich in limestone bedrock such as the southern Argolid, in fact much of the country of Greece, karstic development is pervasive. In the area around Franchthi Cave, however, there are few known cavern systems that are connected to the present land surface. In fact, Franchthi cave is the only large, horizontal cavern in a wide area, but other karstic manifestations do occur in the area (see Figure 3.1 and Jacobsen and Kopper 1981). In the Franchthi headland itself there are two deep karstic shafts, one on the top of the headland downslope from the altitude marker ("160.10" on Figure 3.1) that descends more than 30 m vertically, and another that was discovered in quarry operations at the southeast end of the headland. The latter is known to descend as deep as 104 m below sea level. It is unknown whether either of these shafts is connected in any way to Franchthi Cave.

On a smaller scale, there are numerous karstic springs in the vicinity of the coastline below the Franchthi headland (van Andel and Sutton 1987:18 and Figure 5; Jameson et al. 1994:169–172). Most of these are submerged at present, some just below sea level (Figure 3.2) and others as deep as −180 m, but some of them were accessible until the Bronze Age. Many of these spring vents are situated along the contact between the Franchthi limestone and the underlying flysch or ophiolite.

A little farther away—about 5 km northeast of Franchthi Cave—is a large karstic basin, or polje, in which the modern town of Dhidhima sits (van Andel and Sutton 1987: Figure 1). A large sinkhole occurs on the limestone flanks above the alluvial surface of the polje. In the floor of the alluvial plain itself is another sinkhole that exposes the extensive alluvial fill of the polje. Another large sinkhole is located near Iliokastro, about 13 km ENE of Franchthi Cave, which might be connected to a significant underground gallery, but the bottom of the sinkhole is impenetrable, being filled with collapsed rock debris at present.

This is an incomplete inventory of karstic features in the Franchthi area, but provides a background for the discussion of Franchthi Cave itself. More details are available in Jameson et al. (1994).

CONFIGURATION OF FRANCHTHI CAVE

General

The habitable part of Franchthi Cave is about 150 m long and ranges from about 30 m wide at the entrance to about 45 m at the widest part in the center of the cave (Jacobsen and Farrand 1987:Plate 2). The cave has two large "windows" to the sky—areas where the roof of the cave has collapsed, producing mounds of breakdown debris on the cave floor (Figure 3.2).

Breakdown

The largest window is about midway from the entrance to the rear of the cave and has an opening of about 40 m diameter (Figure 3.2, Plate 1b). It is underlain by gigantic blocks of limestone, one of which was estimated to be 764 m³ in volume and to weigh about 1835 metric tons (Jacobsen and Kopper 1981:89), in a chaotic arrangement suggesting one large collapse event (Plates 3b, 4a). The summit of this rock pile is 31.8 m above sea level, or about 20 m above the level of the currently habitable area in the front part of the cave. Some mounds of dripstone ("cave coral") have formed on the top and over the edges of some of these blocks (Plates 3b, 4a), indicating some antiquity, but they have not been dated. It appears that this collapse occurred at or near the end of Neolithic occupation of the cave. Neolithic occupational layers in the outer part of the cave do not lap onto this rock pile, and a small area of Neolithic occupation was discovered under the innermost edge of the rock pile (Notebook 560:7, 1974) about at the point marked "1974" on the cave plan (Figure 2.1) at an elevation of 8.9 m above sea level. A small sondage here revealed mixed Neolithic sherds; the earlier sherds were "more heavily worn and encrusted" than those of the Final Neolithic (FCP 5), according to Vitelli (1999).

The second, smaller window is located at the inner end of the cave (Figure 3.2, Plate 3b). It is about 20 m long and about half that wide, its length being parallel to the strike of the limestone stratification in that part of the cave. The breakdown blocks are much smaller and the rock pile less voluminous than that under the big window, and there is no notable accumulation of cave-floor sediments in this area. No excavation took place in this area, so it is unknown what underlies this rock pile. Kopper considered this window to be a fault opening or "slump" of recent age, perhaps a fault the offset of which conceals a continuation of the cave to the southeast under the headland (Jacobsen and Kopper 1981:98). I think this interpretation is not very likely. There is no obvious trace of such a fault across the top of the headland, where one might expect one, if the fault is so recent. Furthermore, minor slippage along bedding planes in this area where the stratification is nearly vertical (Plate 2a) is not unexpected and could give the illusion of fault movement (cf. Jacobsen and Kopper 1981:Figure 8).

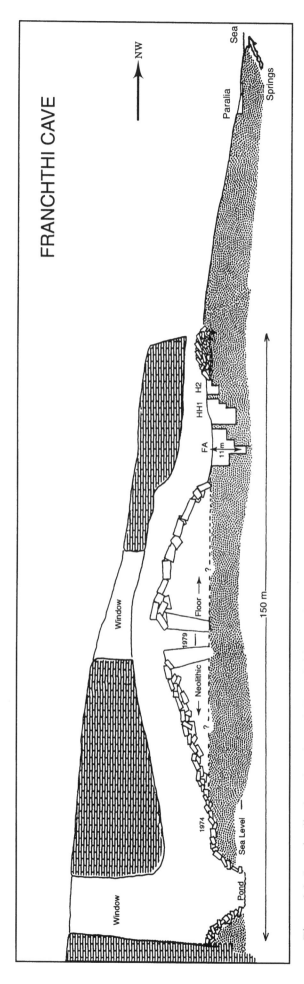

Figure 3.2. Longitudinal section through Franchthi Cave, in part schematic. The figures 1974 and 1979 mark the positions of the Neolithic sondage toward the rear of the cave and of the prominent chasm among the gigantic limestone blocks, respectively. Note also the area of Paralia excavations and the freshwater springs along the coastline (right). Modified from Farrand (1993, Figure 1).

Another area of collapse is found under the brow at the present cave entrance (Plates 2a, 3a). The pile of breakdown debris here also appears to be geologically recent. Trench H2 was excavated at the inside edge of this rock pile, as can be seen in Figures 2.1 and 3.3 and Plate 3a. The excavation revealed that some of the limestone blocks overlay the topmost Neolithic levels of Trench H2, and others seemed to be coeval with Neolithic occupation. Along the north edge of H2, Neolithic levels lapped onto some of the large blocks of the rock pile. Unit H2B:31, about 0.9 m below datum, was the first excavation unit to pass under the lowest of the "great stones." According to excavators, the potsherds from H2B:31 were mid-MN types, and Vitelli (1993:64) places all units of H2B in FCP 2.2, dating this rock fall to a geological event during the MN occupation of the cave.

Still older collapse of the brow can be inferred from Trenches B through E outside the present cave entrance, as mentioned in Chapter 2. At depths of <1 m, Neolithic artifacts of FCP 1, as documented in Trench BE Balk (Vitelli 1993:42), overlie large limestone blocks that were too consolidated to be excavated.

The Cave Floor

The floor of the cave, thus, is broken into distinct areas defined and limited by the large piles of breakdown debris. The presently habitable area lies between the breakdown under the brow and the forward edge of the tremendous rock pile under the central window. This 35 by 38 m area is not uniformly level. Much of it lies about 11 m above sea level, but there is a low area on the central west side that descends a bit below 10 m—the area of Trench FA—and a high area, just above 12 m, extending inward from the rock pile under the brow into the area of Trenches H2 and HH1(see Figure 2.1 and Plate 3a).

The topography of the cave floor in the area of our excavations has changed somewhat over the past 30,000 years or so. At times in the prehistoric past, the low area in the vicinity of Trench FA was even lower relative to the floor area around Trench HH1 than it is today. See Table 7.1 and discussion in Chapter 7.

After climbing over the central rock pile towards the rear of the cave, one descends nearly to sea level in the area of a small pool of standing water, but there is very little level surface in the back of the cave. From the pond, one climbs up again into the area under the innermost window. As mentioned above, a small area of Neolithic deposits was discovered in the rear of the large breakdown pile at about 8.9 m above sea level, or at about the same level as Neolithic deposits in the front of the cave. This suggests the presence of a more or less level floor throughout most of the cave prior to the collapse under the big window.

The Unseen Cave

The accessible area within Franchthi Cave is part of a more extensive karstic network, of which very little is known. The pool, mentioned above, provided a glimpse of this larger cavern system. Only a small portion of the pool (about 30 m long and 4 m wide) extends into the open cave, but one can see it disappear under the southeastern cave wall. The pond was investigated by divers Steve Hallin and Ralph Mason along with James Dengate in 1973. Their observations (Notebook 542) are summarized in the following paragraphs, and a more complete narrative by Dengate appears in Vitelli (1999).

The bottom of the pool drops off irregularly from about 2 m depth at 5 m offshore to about 13 m at 40 m offshore, then levels off as far as measurements were taken, about 50 m offshore. Beyond this point the cave passage "opens into a gigantic room [estimated to be about 10 by 12 m diameter] with stalactites and stalagmites" up to an estimated 0.5 m diameter. Some are joined into columns. The submerged cave appears to extend, or connect with another cavern, to the SW under the area of the innermost window chamber.

Abundant potsherds and bones were concentrated near the pool edge. The divers' notes indicate that "a large amount of objects" [artifacts?] covered the bottom "from the point of entry all the way to the point of deepest penetration" (Notebook 542), i.e., as far as 45 m offshore and about 13.5 m depth, but Dengate (in Vitelli, 1999) says that artifacts "ceased entirely after about 20 m from the pool edge." Final Neolithic sherds were predominant, but about 28% of the sherds were mostly Archaic, Classical, and late Roman to early Byzantine, according to Dengate (in Vitelli, 1999). Also recent, or very young pieces of wood were recovered. There is potential for older archaeological materials to occur under the floor of the pond, given that it was presumably a dry cavity when sea level was much lower than present during Upper Palaeolithic, Mesolithic, and early Neolithic times. On the other hand, the pond must have existed in more or less its present configuration in Classical and Roman times.

The floor of the pond is covered by fine red silt, which becomes increasingly thick in deeper water. The silt is said to be "humus-like" and is very easily disturbed, obscuring the divers' vision. Angular limestone fragments 10 to 20 cm diameter project upward through the silt where the silt is thin, and some large limestone blocks, 3 to 5 m in diameter, sit on the pond floor. Limestone clasts from the fragmented cave wall were easily detached by air bubbles released by the divers.

No movement or circulation of the water was observed. The water was very clear and was stratified into a surface layer, variously estimated to be from 0.3 to 1 m thick, that was colder and much less salty than the lower layer, which was characterized as "very salty." Salinity measurements by Riley Schaeffer (formerly of the Chemistry Department, Indiana University, now Professor Emeritus at the University of New Mexico), expressed as chlorosity (in milligrams of chlorine/liter, or mg Cl/l), were 6.7 at the surface, but increased to as much as 15.3 mg Cl/l at -1.5 m and was always >10.7 mg Cl/l at greater depths. By comparison, worldwide sea water is about 19 mg Cl/l, and the Mediterranean is even saltier, according to Schaeffer. Additional details are given by Dengate (in Vitelli, 1999).

Through the years of the Franchthi excavations, eels were observed in the pond, and one was killed by the divers in 1973 because it was "in the way of further exploration." The presence of eels suggests that there must be a connection between the pond and the waters of Kiladha Bay in one direction or another.

THE SEDIMENTARY FILL

The sedimentary fill in Franchthi Cave has been observed over only about 10% of the area of the outer, presently habitable part of the cave. (See also Chapter 2.) The greatest observed thickness of sediments is in Trench FA, from ca. 10 m above to ca. 1 m below the Greek sea-level datum, thus about 11 m thick. Electric resistivity measurements indicate a considerably greater sediment thickness, however. With the greatest possible electrode spacing, equivalent to a depth of about 16.5 m, the resistivity soundings did not reach the bedrock floor in the outer part of the cave (Jacobsen and Kopper 1981:104). Thus, the bedrock floor in the area of our excavations must be more than 5 m below sea level. Such a depth is also suggested by the depth of the pond in the back of the cave, the floor of which lies as much as 13 m below sea level (see above). Thus, there is potential for a considerable thickness of archaeological deposits below those reached in the excavated trenches. Recall that during cave occupation sea level was at times as low as 100 to 120 m below present sea level and never higher than ca. -7 m, even at 5000 B.P. (Jameson et al. 1994:200–201).

There is also some archaeological evidence of older artifact-bearing levels below those reached in our excavations. Mousterian points and pieces with Levallois preparation have been found in disturbed and mixed deposits that overlie intact Neolithic levels, and the lowest units excavated in Trench FA show indications of lithics and fauna of possible Middle Palaeolithic affinity (Perlès 1987:49–51). Moreover, the tephra in the lowest levels of Trenches FA and HH1 (described in Chapter 5) is likely to be between 33,000 and 40,000 years old (see dating in Chapter 6), which means that its age is close to the normal limit between Middle and Upper Palaeolithic industries in Europe.

CHAPTER FOUR

Lithostratigraphy

INTRODUCTION

General

A lithostratigraphic unit is defined as a body of sediment more or less uniform in its physical characteristics and different from underlying, overlying, or laterally adjacent sediment bodies. The degree of uniformity or the allowable amount of internal variability within a lithostratigraphic unit is commonly a judgment call by the stratigrapher and may vary according to the scale or goals of the field operations. For example, the entire sedimentary fill of a cave such as Franchthi may constitute a single lithostratigraphic unit for a stratigrapher who is mapping the bedrock geology of an area. On the other hand, that fill could be subdivided into tens or even hundreds of strata, if one is working on a very fine scale. Hypothetically, each of our excavation units could be a lithostratigraphic unit if it were indeed physically different from the adjacent units.

Lithostratigraphic units are also characterized by the nature of the bounding surfaces (called "contacts" in geology) that separate one from the next. These contacts may vary from very sharp to very diffuse, and may be planar, wavy, or irregular. We have tried to make those distinctions on the primary section drawings by the use of various line symbols and weights (Jacobsen and Farrand 1987:18). In some cases, these boundary conditions are mentioned below.

Lithostratigraphy in Franchthi Cave

The lithostratigraphic sequence for Franchthi has been derived without recourse to biostratigraphic or ethnostratigraphic zonations of my colleagues in order to be as objective as possible. See Jacobsen and Farrand (1987:Figure 5) for a brief, theoretical discussion of multiple stratigraphic hierarchies, and a preliminary comparison of different hierarchies in Franchthi Cave appears in Farrand (1993:Figure 6). A complete discussion of the multiple stratigraphies will appear in our synthetic volume to be published later. The sequence begins at the surface with Stratum Z and continues in reverse alphabetical order with depth, as is the practice in geological stratigraphy. Some units have been subdivided by adding numbers (3, 2, 1) to the stratum letter, and some have been called "members" and others are given "stratum" status. Although those differences are to some extent arbitrary, they are intended to indicate that members have similarities or coherence that imply that they are parts of a single sedimentary episode, e.g., Members W1, W2, and W3, and Members T1, T2, and T3. In contrast, Strata Y1, Y2, and Y3 have contrasting sediment types. One could argue for more logical or more numerous subdivisions (as mentioned above), but the designations used here were used in previous publications (Farrand 1993), except for the addition of Stratum Y3. There is no theoretical reason to change them, and changing them at this late date would likely create confusion. *The stratum and member designations are merely names of lithological entities, and they carry with them no implications regarding the thickness, importance, duration, or cultural identity of the strata.*

The lithostratigraphic units in Franchthi Cave were determined by this writer after the conclusion of the excavations. They were not developed or used during the course of the excavations. They are based on the section drawings published in Fascicle 1 (Jacobsen and Farrand 1987) as verified in the field in 1979 and amplified by excavators' descriptions in the field notebooks. In addition, the schematic section drawings (also in Jacobsen and Farrand 1987) were overlaid on the section drawings to clarify ambiguities and to supply information not available from the drawings or notebooks.

A pervasive problem was the placement of "scarp tags" on the section drawings. (The section drawings were not always completed immediately after a trench was freshly excavated, and in some cases not until the following season.) Scarp tags with the excavation unit number were initially placed by the excavators at the base of each unit as it was completed. Over the course of the years of excavation, some tags fell out and some became illegible so that they could not be used during verification in 1979. Other tags had to be moved by the excavators in the course of excavating a later trench that removed the adjacent wall of an earlier trench. They replaced the tags to the best of their judgment, but not accurately in all cases, and a number of tags were lost in this process. Thus, the position of a numbered scarp tag on a section drawing is not necessarily definitive.

The schematic section drawings have their limitations, too. They are based on depths below datum taken only in the four corners of the trench—or as near to the corner as practicable. Thus, the straight lines indicating the top and bottom of an excavation unit on a schematic drawing are not necessarily accurate depictions of the topography of that unit except in the corners. Thus, it may appear, when overlaying a schematic drawing on a section drawing, that a unit crosscut natural stratigraphy when that was not really the case. Nevertheless, the schematic drawings along with the notebook descriptions were very useful in interpreting the apparent lithostratigraphy shown on the sections drawings.

Correlation

The following descriptions include the excavation units that are considered to be correlative from trench to trench based on their lithological and sequential characteristics. To the extent possible these correlations are based on the macroscopic physical features of lithological units in the different trenches, such as relatively thin but very rocky layers (Strata V and X1) and the unique volcanic tephra (Stratum Q). Radiocarbon dates were used to corroborate the correlations and to sort out ambiguities, but they were never the primary basis for lithostratigraphic correlation. (See "Intertrench Correlation" in Chapter 6.) In no case were the laboratory analyses used as primary means of correlation in order to avoid circular arguments in the interpretation of the sedimentological history of the sequence in each trench (Chapter 5).

COLOR DETERMINATIONS

Sediment colors have been standardized to the extent possible according to the Munsell Soil Color Charts (1975 or other editions). At Franchthi Cave, no standardized color chart was used in the first season in 1967. In 1968 excavators began the use of the Geological Society of

America Rock Color Chart (1991), which is based on the Munsell system. Only in 1971 did the Munsell Soil Color Chart begin to be used. Nevertheless, a number of problems and ambiguities arose across the eight seasons of excavations. In the first place, it is apparent upon cross-checking various readings that some of the excavators were not skilled in the use of the color charts and did not recognize the subtle distinctions that are possible with the Munsell system. Moreover, the excavators rarely noted the moisture content of the sediment that they were coding, which can change the "Value" of the Munsell code by 1 to 3 units, e.g., from 10YR 3/3 (dry) to 10YR 6/3 (moist) for a given sample, corresponding to "dark brown" and "pale brown," respectively.

Ambiguities also stemmed from the use of the *rock* color chart, relative to the *soil* color chart. The rock color chart contains fewer pages, omitting the intermediate 2.5 and 7.5 hues, and it contains many fewer color chips on each page, necessitating interpolations between those chips present. From the notebooks it is apparent that the excavators did not make such interpolations, but simply designated the chip nearest to the sediment color, according to their judgment. Moreover, unfortunately, the verbal color names used in the rock color chart are not the same in some cases as those used in the more detailed soil color chart, e.g., the color code 5YR 5/6 is called "light brown" in the rock chart, but "yellowish red" in the soil chart. Even more confusing, "moderate brown" on the rock chart, a color commonly cited by the excavators, may be either 5YR 4/4 or 5YR 3/4, that is, "reddish brown" or "dark reddish brown," respectively, on the soil color chart. To complicate matters further, excavators using the rock color chart usually wrote down only the verbal name for the color, and not the Munsell code, so that one does not know whether "moderate brown" meant 5YR 4/4 or 3/4, or whether "light brown" meant 5YR 6/4 or 5/6.

These ambiguities remain, and in the following descriptions it will be noted that the colors given for my sediment samples (in the last paragraph for each stratum) commonly differ from those given in the notebooks that are mentioned earlier in each description. The colors of the sediment samples were observed under laboratory conditions and determined by consensus of three or four persons experienced in the use of the Munsell system.

LITHOSTRATIGRAPHIC DESCRIPTIONS

The lithostratigraphic units (strata) are illustrated on Figures 4.1 for FA and 4.2 for HH1. In the descriptions below, the following format is used:

- Major or key *excavation units* included in the stratum. A complete listing of all excavation units organized by stratum is given in Table 4.1.
- Key uncalibrated *radiocarbon dates* from the stratum. Only the uncalibrated dates are given here for simplicity. All the dates and their calibrated equivalents are found in Table 6.1.
- One or two paragraphs of *description gleaned from the excavators' notebooks.* This information is rather uneven because different excavators were involved from year to year, and variable manners of expression were used by excavators with different backgrounds. In some cases it was a challenge to "translate" the excavators' comments into geological terminology.
- A description of the *sediment samples* collected from each stratum, including the Munsell color, granulometric texture, and mention of artifact and faunal content. The full description of these samples is given in Chapter 5, especially in Figure

5.2. A complete list of sediment samples in Appendix A includes the excavation units represented by each sample, its depth, thickness, initial weight, moist and dry Munsell colors, and the field description.

Stratum Z

Units A:1–20 and top of 40; F:1–4; F1:1–6; FA:1–45:
Units H:1–12; H:29–31; H1:1–39; H1:41,44,47; H2A1–19:

In A–FA–FF1 pale to dark yellowish brown sediments with hearths, ash, and charcoal—one sample dated at 105 B.P. (A.D. 1845)—and scattered large rocks, up to 50 to 60 cm across. The thickness of Stratum Z is quite variable.

In HH1 and H2, light grayish brown, fine-grained rubble with loamy matrix (Plate 4b); small rock fragments more or less randomly distributed throughout, mostly 2 to 4 cm, and standing at high, odd angles, although they tend to lie horizontally in some places; rarely is there any suggestion of stratification, giving an overall impression of homogeneity; includes occasional lenses of ash and charcoal, and small lenses or pods of bright reddish brown (pinkish) silt, usually free of stones, and commonly 2 to 5 cm thick.

Wheel-thrown pottery, iron nails, bottle glass, and modern wood persist alongside Neolithic sherds at least as deep as −1.88 m below the surface in Trenches H1 and H1 Terrace, and down to −2.23 m in Trench H, according to the excavators' notes. (See also Vitelli 1993.) In addition, several fragments of copper were recovered at −2.82 m in Trench H1

Total thickness in HH1 nearly 2 m; lower contact very irregular, filling hollows (perhaps modern pits) in Stratum Y2 up to 1 m deep; upper surface is present floor of the cave.

Stratum Y3

Recognized in Trench A–FA–FF1 only. Units A:21–26 and lower 40; FAS:59–74; FAN:59–63, 70–72, 75, 78, 79; FF1:7–10; FA:46–49:
Dates: 5160 (FA:39) to 5260 B.P. (FAS:72)

Pale to dark yellowish brown with hearths, ash, and charcoal (Plate 7a). An especially large hearth complex occurs in FAS:73,78,79. Large (Neolithic) pits in east part of FAN, e.g., FAN: 65 and 68, and adjacent part of FA East Balk. Maximum thickness about 1 m.

Sediment sample FR 1-1 (Units FA:70 and 74) is stony, very dark brown (10YR 2/6, dry) loam.

Stratum Y2

Units A:28–39, 41–58; FAS:84–127; FAN:73, 76,85–137; FF1:11–42B1 and 40A1; FA:49A–57; H1:40–71; H1B:72–74; H:13–28, 32–39; H37P , H2A:20–134.
Dates: 6110 (FAN:89) through 6670 B.P. (FAN:137); 6750 B.P. (FF1:18,19); 6830 B.P. (FF1:34).

A thick stratum in Trench FA, ca. 2 m (Plate 7a), but only ca. 1 m in HH1. Generally grayish to yellowish brown, mostly 10YR 5/3, 5/4, and 6/3, moist, but with somewhat reddish hues. Variable stoniness, mostly smaller rock fragments, but some horizons with rocks up to 30 to 35 cm in lenses or clusters, e.g., FAS:98–102, FAS:118, FAN:110, FAN:131, at least some of which are associated with hearths or hearth complexes, e.g., FAS:98–102.

For Trench HH1 Stratum Y2 is not well depicted on the section drawings. In general, it is pale reddish brown, loose, coarse-grained rubble (*éboulis secs*), with sharp, angular rock fragments 5 to >30 cm diameter and containing scattered, mixed sherds. Stratum Y2 here contained a number of large limestone blocks, up to ca. 80 cm, as shown on Figure 4.2 and Plate 4b.

Table 4.1. Correlation of Excavation Units with Lithostratigraphic Units

STRATUM	H2	H PED	HB	H & HA	H1A	H1 & H1B	G & G1	FA BALK,EB	FA:QSW,WB	FF1	FAN	FAS	A	STRATUM
Z	H2A:1-19			H:1-12		H1:1-39	disturbed	FA:1-22	FA:23-29	F:1-4	n.a.	n.a.	A:1-20	Z
				H:29-31		H1:41,44,47		FA:30,33A	FA:31-33B	F1:1-6			A:40 top	
								FA:34-36	FA:37-40A,45					
								FA:41-44						
Y3	absent	absent	absent	absent	absent	absent	disturbed	FA:46-49	not dug	FF1:7-10	FAN:59-63	FAS:59-74	A:21-26	Y3
									below here		FAN:70-72		A:40 bottom	
											FAN:75,78,79			
x cut											FAN:64-68	FAS:76-83	?	x cut
											FAN:82-84,88			
Y2	H2A:20-134	H:37A-37π		H:13-28	absent	H1:40-71	disturbed	FA:49A-57		FF1:11 thru	FAN:73,76	FAS:84-126	A:28-39	Y2
				H:32-39		except 41,44, 47				FF1:42B1	FAN:85-142		A:41-58	
						H1B:72-74				& 40A1				
x cut								not dug below here			FAN:138-141	FAS:128		x cut
Y1	H2A:136-148	H:37P-EE		H:40	H1A:72-73	H1B:75-79	G:25,26			FF1:43B1-B2	FAN:143-146	FAS:128-133	A:59-60	Y1
	H2APed:180-188						G1:19			FF1:40A2-41A1				
x cut											FAN:148-149	FAS:134,135?		x cut
X2	H23A:150-153	not dug		H:41-49	H1A:74-108	H12B:80-114	G:27-34			FF1:41A2-A4	FAN:150-160	FAS:136-145	A:61-64	X2
	H2APed:189-197	below here					G1:20-27			FF1:43A1				
										FF1:44B1-B4				
										FF1:40D-42D				
x cut				H:50(w)	H1A:104	H1B:115					FAN:161-163	FAS:146		x cut
					H1A:110-113									
X1	missing?			H:50(w)	H1A:110-112(w)	H1B:115	missing?			FF1:44B5 -	FAN:170-172	FAS:148-150	A:65	X1
				H:46-47(e)	H1A:105-115(e)					FF1:45B4				
x cut				H:47	H1A:113,114	H1B:117,118					FAN:170	FAS:151		x cut

Table 4.1. (Cont.)

STRATUM	H2	H PED	HB	H & HA	H1A	H1 & H1B	G & G1	FA BALK,EB	FA:QSW,WB	FF1	FAN	FAS	A	STRATUM
					H1A:117						FAN:172-173			
W3	H2A:150-163			H:49,50(e)	H1A:117-131	H1B:117-127				not dug below	FAN:174-184	FAS:152-159	A:66	W3
	H2APed:197-203			H:51(w)										
x cut				H:51,52		H1B:128,129					FAN:186	FAS:160,161		x cut
												FAS:162?		
W2	not dug below here		H:61B1-	H:53A1 -	H1A:132-155	H1B:128-141	G1:38-59				FAN:187-230	FAS:163-192	A:67	W2
			H:63B4	H:54A7										
x cut			H:63B4	H:54A	H1A:154,155	H1B:141					not dug below	FAS:195-202	not dug below	x cut
W1			H:63B4 -	H:54A5 -	H1A:155-160	H1B:141-147	G1:60-65					FAS:196,197		W1
			H:65B	H:55A3										
x cut			H:65B	H:55A	H1A:162-168	H1B:148,149						FAS:195-202		x cut
V			H:65B	H:55A2-A4(e)	H1A:161-168	H1B:148-150	G1:66,67					FAS:199-202		V
				H:56A1, A2(w)										
x cut			H:66B	H:56A2-A3	H1A:167	H1B:151	not dug below here					FAS:197-202		x cut
U			H:67B	H:56A3 -	H1A:169-175	H1B:151-154						FAS:203-206		U
				H:57A										
x cut			H:67B5,6	H:56A4 -	H1A:174-176	H1B:154,155						FAS:207		x cut
				H:57A3										
T3			H:69B,70B	H:58A1-A5	H1A:175-180	H1B:155,156						absent		T3
x cut					H1A:180	H1B:156-157						absent		x cut
T2			H:71B1-4	H:58A6 -	H1A:181-188	H1B:156-159(w)						absent		T2
				H:59A1										

Table 4.1. (Cont.)

STRATUM	A	FAS	FAN	FF1	FA:QSW,WB	FA BALK,EB	G & G1	H1 & H1B	H1A	H & HA	HB	H PED	H2	STRATUM
x cut		absent						H1B:157,160	H1A:186,188		not dug below			x cut
T1		absent						H1B:158, 159(e)	H1A:182-187(e)	H:59A2 -				T1
								H1B:160-166(w)	H1A:187-189(w)	H:60A				
x cut		absent						H1B:159(e)	H1A:188,189					x cut
								H1B:165,166						
S2		absent						H1B:166-172(w)	H1A:186-204(e)	not dug below				S2
								H1B:159-172(e)	H1A:190-204(w)					
x cut								H1B:172						x cut
S1		FAS:208						H1B:173-180	H1A:205-210					S1
x cut									H1A:211					x cut
R		FAS:210-217						H1B:181-212	H1A:212-220					R
Q		FAS:218-222						H1B:213	not dug					Q
x cut		FAS:218												x cut
		FAS:220-223												
P		FAS:224-227						H1B:214-215						P

NOTES:

- Repetition of unit numbers are generally cleaning units.
- Repetition of unit numbers in stratum and cross-cut rows indicates:
  a) general coincidence of the excavation unit(s) with the designated stratum;
  b) deviation (cross-cutting) of an excavation unit above and/or below the stratum in some part of a given trench.
- Unit numbers followed by (e) or (w) indicate different relations on the east and/or west sides of the trench as shown on section drawings.

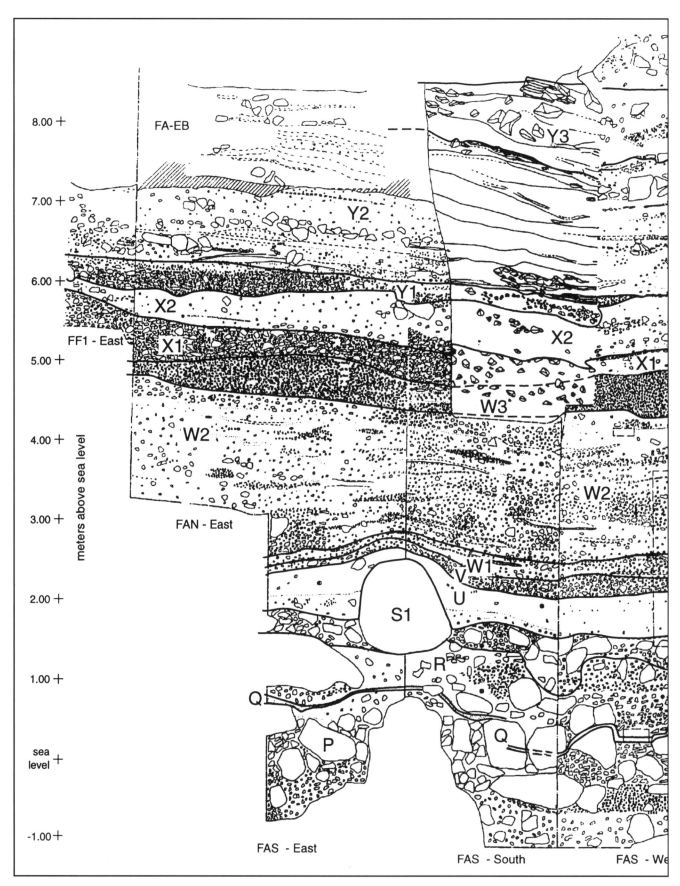

Figure 4.1. Montage of section drawing of the four sides of the Trench FA-FF1 complex, assembled from Plates 7, 8, 9, 10, 11, 12, 18 in Jacobsen and Farrand (1987), which should be consulted for clarity of details. The positions of

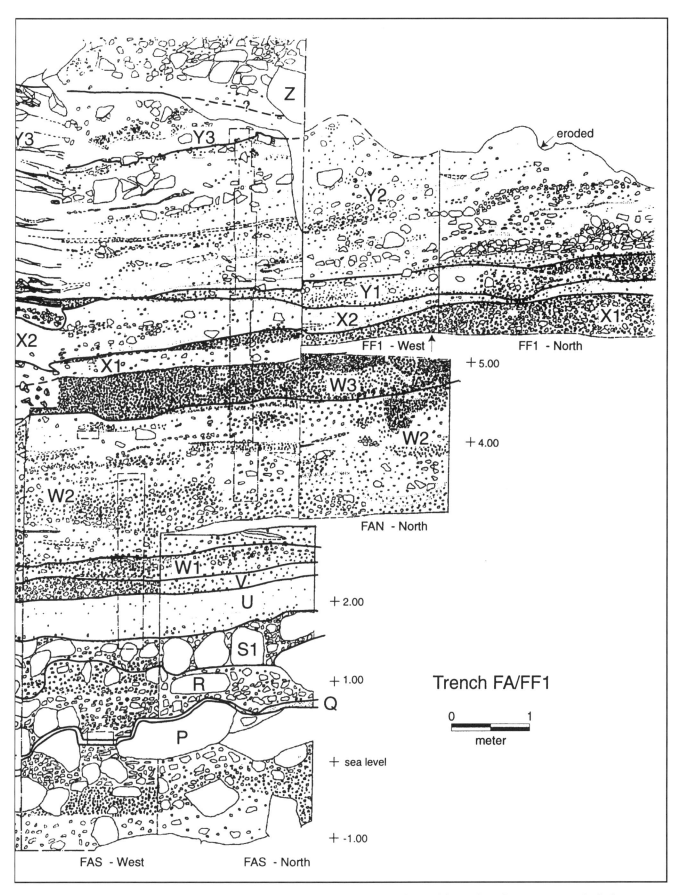

lithostratigraphic units are indicated by the heavy lines across the section. This section does not extend all the way to the present cave floor.

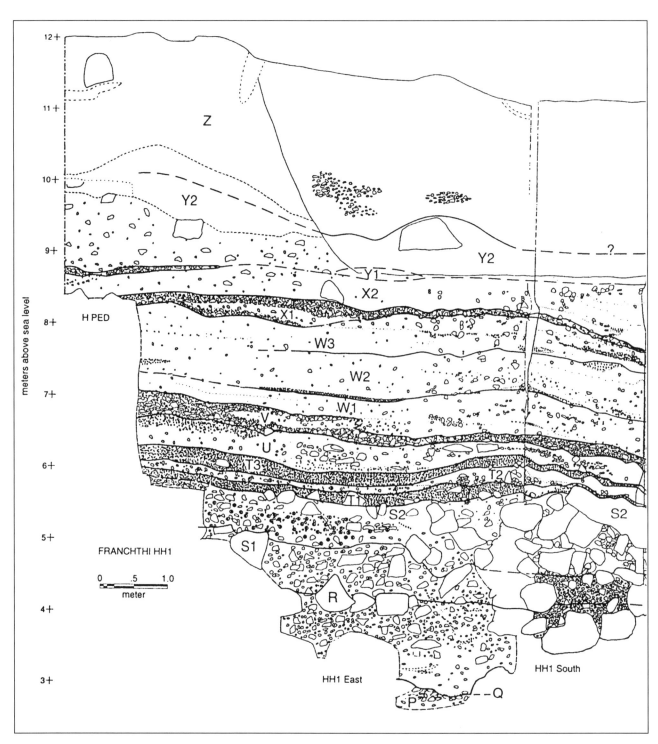

Figure 4.2. Montage of section drawings of the four sides of the Trench HH1 complex, assembled from Plates 13, 14, 15, and 17 of Jacobsen and Farrand (1987), which should be consulted for clarity of details. The top

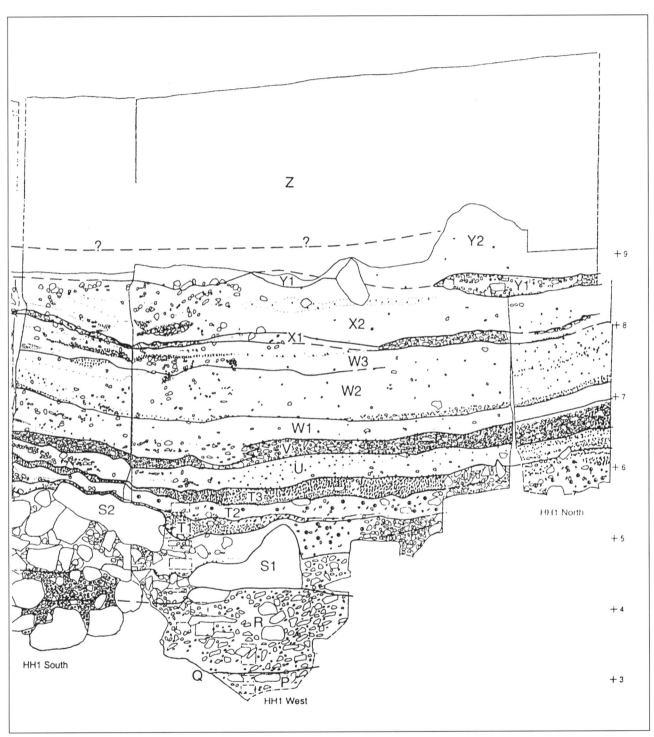

of the section is the present cave floor. The positions of lithostratigraphic units are indicated by the heavy lines across the section.

The large rocks appear to be part of the same rock fall exposed in Trench H2, which is seen again on the north wall of Trench H, on the south wall of H1 Terrace, and underlying the cemented travertine mass in west-central Trench H. Stratum Y2 was characterized by the excavators as definitely more reddish than the overlying sediments of Stratum Z, although their color codes do not reflect this, apparently subtle, difference.

Sediment samples FR 1–2 through 1–12 from Stratum Y2, all from Trench FAN, are dominantly grayish brown to brown loams (commonly 10YR 5/2, 5/3, 6/3, dry) with variable stoniness.

Stratum Y1

Units A:59–60; FAS:129–133; FAN:142–146; FF1:43B1, 2 and 40A2–41A1; G:25, 26; GG1:19; H1B:75–79; H1A:72–73; H:40; H:37P–EE; H2A:136–148; H2A Ped: 180–188.
Dates: 6940 (FAS:129), 7280(?) (H:37Y), and possibly 7190 B.P. (A:56).

A thin to discontinuous rocky layer, called "rocky red" by excavators (Plate 8b), thickest in A-FA-FF1 (up to 40 cm, but thins to a few cm on the west side of FAS); represented by discontinuous lenses in HH1, e.g., in H Ped (H:37P–EE) and H1A:72 and 73, as best can be reconstructed from the sketchy section drawings and notes. FAS:131–133 was characterized as "red gritty" sediment by excavators, which was confirmed by sediment sample FR 1–15 from adjacent FAN:146. Rock-fragment sizes not specified, but are dominantly 2–32 mm gravel in the sediment sample.

Sediment samples FR 1–14 and 15 were pale brown (10 YR 6/3, dry) to yellowish brown (10YR 5/4, dry) stony loam to sandy gravel.

Stratum X2

Units A:61–64; FAS:136–145; FAN:150–160; FF1:41A2–A4, 44B1–B4, 40D–42D; G:27–34; G1:20–27; H1B:80–114; H1A:74–108; H:41–49; H2A:150–151; H2A Ped:189–197.
Dates: 7790 (A:63), 7900 (FF1:44B5), 7930 (FAS:143), 7980 B.P. (FAS:146), and 8020 B.P. (FF1:43A1).

A moderately thick stratum, 35 to 60 cm in A-FA-FF1 and 50 to 80 cm in HH1, that apparently accumulated rapidly, but is relatively free of stones. However, rounded "sea" pebbles were noted in units FAS:136, 143, and 145, and in H1A:85. Stratum X2 was characterized as a "gray clay" stratum by excavators in Trench FA (Plate 8c, upper half), although their Munsell Soil Chart colors are generally brown, dark brown, and dark yellowish brown; in HH1 the excavators generally called the colors "brown" and coded them as "light moderate brown," probably equivalent to 5YR 5/4, which is light reddish brown on the soil color chart. The "gray clay" stratum is generally congruent with Vitelli's (1993) Franchthi Ceramic Phase 0/1, although many individual excavation units in this stratum yielded no sherds at all.

Sediment samples FR 1–18 and 19 are pale brown (10YR 6/3, dry) variably stony silt loam with little cultural debris.

Stratum X1

Units A:65; FAS:148–150; FAN:163–172; FF1:45B1–B4; H1B:115; H1A:110–115; H:50 and 46–47; missing in H2A
Date: none

Another "rocky red" stratum, in fact *éboulis secs* (see Chapter 5 and Figure 5.1), thin but conspicuous in HH1 (Plate 5a, at top), and up to 40 to 60 cm thick in A-FA-FF1 (Plate 8c, lower half), constituting a good marker bed for correlation from trench to trench. Munsell soil colors are mostly 5YR 3/2-3/4, dark reddish brown, to 7.5YR 4/4, dark brown, according to excavators. In addition to abundant stones, this stratum is rich in crushed land snails and also contains marine shells.

Sediment samples FR 1–20 to 1–22 are brown to yellowish brown (10YR 5/3 and 10YR 5/4, dry) sandy loams with abundant stones. Rock fragments between 128 and 32 mm constitute up to 60% of each sample, and many of these rocks are punky and have a whitish surface alteration, as do some of the snail shells, indicative of weathering. A few rounded pebbles appeared in sample FR 1–21 (units FAN:165, 167). The sharp upper contact (Plate 8c) and the altered rocks suggest a definite unconformity at the top of this unit that may mark the Neolithic/Mesolithic boundary.

Stratum W

Stratum W has three members W3, W2, and W1, from top to bottom.

Member W3

Units A:66; FAS:152–159; FAN:174–184; G:35–42; G1:28–37; H1B:117–127; H1A:117–131; H:49–51; H2A:152–163; H2A Ped:197–203.
Dates: 8530(?), 8710, and 8730 B.P., all from FAN:177.

W3 is the upper member of a very thick Stratum W, characterized by land snail middens, much ash, and charcoal. "Polished sea pebbles" were observed in units FAS:151 and 157. Member W3 is particularly rich in snail remains and is rather rocky as well (Plate 8d); it is brown/dark brown (7.5YR 4/4) to dark reddish brown (5YR 3/4) with piles, "humps," and pits filled with crushed snail shells, along with fire-cracked rocks. On the whole this stratum is not as stony as X1, but it appears to be transitional with X1. Member W3 is generally 50 to 60 cm thick in both A-FA-FF1 and HH1.

Sediment samples FR 1–23 (FAN:172–174) and FR 1–24 (FAN:179–180) are yellowish brown (10YR 5/4, dry) silty loam with moderately abundant stones (50% between 128 and 32 mm), many snails, some bone, and charcoal. Two rounded pebbles were included in FR 1–23.

Member W2

Units A:67; FAS:163–192; FAN:187–230; G1:38–59; H1B:128–141; H1A:132–155; H:53A1–54A7; H:61B1–63B4.
Dates: 9150 (FAN:197), 9250 (FAN:218), 9290 (H1B:126), 9270 and 9300 (H1B:139, 2 dates), and 9300 B.P. (H:61B1).

W2 is the thickest member of Stratum W, especially in A-FA-FF1 where it is rather uniformly 1.6 to 1.75 m thick. In HH1 Member W2 is about 0.6 to 0.7 m in thickness. Radiocarbon dates indicate that this member, and W1 below it, accumulated very quickly, as much as 250 cm per 100 years, or 2.5 cm per year. The character of Member W2 is best seen in A-FA-FF1 on Figure 4.1 where it appears as a complex of small lenses of charcoal and ash, numerous hearths, and snail shells (mostly crushed) set in a matrix of dark brown (10YR 3/2), dark reddish brown (5YR 3/3), and dark grayish brown (10YR 4/2) loams and clay loams, with occasional lenses of "rocky red" sediment (e.g., FAN:196). Some snail shell accumulations appear to be in troughs, according to excavators. Bones, including those of large fish, marine shells, and "polished sea pebbles" occur throughout (notably in unit FAN:198) indicating much anthropogenic influence. Although rock fragments occur throughout, the lower part of Member W2 appears to be stonier—fragments 128 to 32 mm comprise 40 to 50% of the sediment sample—than the middle and upper parts, e.g., FR 1-32 (FAN:224) and FR 2-6 (FAS:192). Lenses of stony, shell-rich sediment occur also at the base of Member W2 in HH1, specifically H1A:148–152, H:54A7, and H:63B4, as seen in Figure 4.2.

Sediment samples FR 1–26 through 1–32, FR 2–0 through 2–6, and FR 3–6 through 3–9 are brown (10YR 5/3, dry) and yellowish brown (10YR 5/4, dry) loams and clay loams with variable stoniness and greater or lesser amounts of crushed snail shells. Moreover, a number of rock fragments in the sediment samples were observed to have highly altered surfaces, that is, they are punky or fragile with white rinds, especially in FAN:219–221 (in an apparent pit) and in the correlative units FAS:185–186, although this condition was not explicitly noted by the FAS excavators.

Member W1

Units   FAS:195–197;   G1:60–65;   H1B:141–149;   H1A:155–160;   H:54A5–55A3; H:63B4–65B.
Dates: 9060 and 9430 (both FAS:195) and 9210 and 9280 B.P. (both FAS:197).

The lowest member of Stratum W is distinguished by a change in sediment type, "the first major soil change" in a long time, according to excavators. The sediment is somewhat redder (5YR 3/4 vs 7.5YR 3/2) and much poorer in anthropogenic components than overlying members of Stratum W, although some snails and hearths persist (Plate 5a, middle). Contrary to the situation in Member W2, Member W1 is thicker in HH1, up to 50 to 60 cm, than in FA, where it is only 10 to 30 cm. (It was not excavated in Trenches A or FF1.) Serious crosscutting of this layer occurred in the southeast corner of Trench FA, where this member (as well as Strata V and U below it) rises sharply above a very large limestone block, as clearly shown in Figure 4.1. Although the excavation units do show a pronounced dip inverse to the usual dip in this trench, the slope of the units is not as great as the true dip of the beds.

In Trench FA Member W1 is a "redder, shelly clay loam" with much evidence of burning (mostly ash) as characterized by the excavators who wrote that there was "little question that a new stratigraphic unit began with [FAS:195]." In Trench HH1, the colors appeared more brown than red, 7.5YR 5/4 and 4/4, and 10YR 3/3, that is "brown" and "dark brown," according to the excavators, and the upper part of Member W1 was "essentially sterile" with hardly any rocks. Artifacts and bones do increase in abundance downward through this stratum, and a number of ash layers were noted. In Trench G1, excavation of this member was largely preoccupied with removal of a burial, which was surrounded by dark gray, wet, muddy sediment (Cullen 1995).

Sediment samples from member W1 (FR 2–6 and 2–7; FR 3–10 and 3–11) are brown to light brown (10YR 5–6.5/3, dry) loams, moderately stony, and contain charcoal, snails, and bones. Rock fragments are mostly angular and unweathered.

Stratum V

Units FAS:199–202, G1:66,67, H1B:148–150; H1A:161–168; H:55A2–A4; H:56A1–2; H:65B.
Dates: None.

The sediment color changes, according to excavators' notes, from predominantly brown to light yellowish brown, e.g., immediately below the hearth in FAS:198, to dark yellowish brown in H1B, and to very dusky red to dark reddish brown in H1A. This stratum is a conspicuous concentration of rock fragments *(éboulis secs),* especially in southwest corner of FAS. The fragments are 5 to 10 cm with little matrix, but with chalky white surfaces; some charcoal and occasional snail shells occur. Stratum V rises up and over the large limestone block in the SE corner of Trench FA, which is part of a major rock fall in Stratum S1. In Trench HH1 Stratum V is very conspicuous and can be traced all around the trench (Plate 6a). It is interrupted by a horizontal, fine-grained lens (H1A:162–163) at the east wall, and the upper half of the original layer appears to have been eroded in H1B (Figure 4.2). This erosion, the white (altered) rocks, and sharp upper contact point clearly to an unconformity at the top of Stratum V.

Sediment samples from Stratum V (FR 2–8 and FR 3–12) vary from light yellowish brown (10YR 6/4 dry) to grayish brown (10YR 5/2 dry) very stony loam to *éboulis secs,* where 50 to 60% of the sample is between 32 and 128 mm (Figure 5.1). Many rock fragments have thick white rinds. There are some large fragments of charcoal and some snails, but few artifacts. This stratum is a distinctive key bed for correlation from Trench FAS to Trench HH1.

Stratum U

> Units FAS:202–206; H1B:151–154; H1A:169–174, 175(?); H:56A3–H:57A; H:67B.
> Dates: 10,260 (FAS:204); 10,840 (FAS:207); 10,790 (FAS:"209"); 10,460 B.P.
> (H1A:173).

Excavators characterized this stratum as a "new deposit" (FAS:202) and a "sediment unlike any other from the cave" (FAS:203); in general, it is a hard, dry, compact ("practically impossible to actually excavate w/ trowel" [NB 551:107]), reddish brown clay without many shells or artifacts, but enclosing conspicuous "white rocks" throughout. Its thickness is uniformly 40 to 50 cm. Matrix color is reported as 5YR 3–4/3–4, 2.5 YR 2/4, 10R 3/4–3/6, thus generally (very) dusky red or (dark) reddish brown. There are numerous small rock fragments throughout, but few larger rocks. A circle of stones was noted in H1A:169 that the excavators suggested was artificial, perhaps a hearth; the stones in the circle were not white like the others in this stratum. The "white rocks," which are also more rounded than most rock fragments in the cave sediments, suggest that a weathering interval succeeded the deposition of Stratum U.

Sediment samples (FR 2–9, 10, 11; FR 3–13) are light yellowish brown to pale brown (10YR 5–6/3.5–4, dry) clay loam with 25 to 35% clay and <25% rock fragments in the 32 to 128-mm class. None of the snail shells recovered has a "normal" (shiny) surface; all are chalky.

Stratum T

Stratum T has three members, T1, T2, and T3. This stratum of conspicuous snail-shell middens is found only in Trench HH1. Its aggregate thickness ranges from ca. 20 to 70 cm.

Member T3

> Units H1B:155,156(east); H1A:175–179; H:58A1–A5; H:67B5–6 (and H:69B–
> 70B?); not present in Trench FAS.
> Date: 10,880 B.P. (H1A:175).

A very rich land-snail-shell midden with whole and crushed shells in very dark reddish brown to black matrix (2.5YR 2–3/4) and including moderate amount of flint and bone fragments (Plate 5b). Some snails appear to be "in heaps." Its thickness is variable, from 5 to ca. 30 cm.

The position of Member T3 in subtrenches HA and HB is unclear from the available information. From the section drawings it appears that the snail midden T3 continues across HA but ends abruptly against a rockier deposit (H:69B and 70B) in HB, which could be Member T2 that is rising toward the mouth of the cave. On the other hand, the possibility cannot be excluded that H:69B and 70B are simply a rockier facies of Member T3 where it interfingers with rock fall from the entrance of the cave.

Sediment samples (FR 3–14 and 3–15?) are yellowish brown (10YR 4.5/4, dry) clay loam with abundant snails and charcoal, and some small rock fragments. Marine limpets also present. All snails have somewhat chalky surfaces.

Member T2

> Units H1B:156(west)–159; H1A:181–185?; H:58A6–59A1; H:69B–70B.
> Dates: 11,240 (H1A:181) and 11,090 B.P. (H:59A1).

A mixed deposit of bones, charcoal, snails, marine shells, and fish bones in a very loose matrix; it looks like a "dump." Its thickness ranges from 5 to 25 cm. Matrix colors range from very dusky red (10R 3/3) to very pale brown (10YR 7/4). Some of the deposit is cemented with $CaCO_3$, and all the snails are chalky. The section drawings show much lower snail-shell density than in Member T3. There is a rough alignment of stones across the trench in H1A:184, "conceivably a windbreak."

Sediment samples (FR 3–15 and FR 4–1) are yellowish brown (10YR 4/5–4/4, dry) loams with abundant small to medium rock fragments (40 to 45% of the fragments are between 32 and 128 mm), and many bones, some land snails, *Patella*, and some charcoal.

Member T1

> Units H1B:158–159 (east) and 160–166 (west); H1A:182–187 (east) and 187–189 (west); H:59A2–A3; (H:71B1–4 ?).
> Dates 11,930 B.P. (H:71B2–3), probably too old, reworked or crosscut into stratum S2.

This lowest member of Stratum T is again a rich snail-shell midden and includes many animal bones and a few large rocks. Again it looks like "a rubbish deposit" (NB 519:85). The loose matrix is dark reddish brown (5YR 3/3), and the deposit ranges from 10 cm to nearly 30 cm in thickness.

There was serious crosscutting of Member T1 and the underlying Stratum S2, as can been seen by comparing the schematic drawings of excavation units (Jacobsen and Farrand 1987: Plates 35 and 36) with the section drawings (Figure 4.2) and the notebooks. For example, Units H1B:161–166 on the west side of the huge rock pile (in S2) show a very dense snail midden on the west wall (scarp) of FAS, while the same units east of the rock pile are barren of snail remains. It is clearly stated in NB 532:52ff that in units FAS:161–166 snails are "absent" or "nonexistent." In trying to follow natural strata, the excavators were obviously hampered by the presence of very large roof-fall blocks, up to 1.5 m across, lying chaotically in Strata S2 and S1. The natural strata, as shown in section drawings, clearly rise up and over the large limestone block in the southwest corner of FAS, such that they lie some 35 to 40 cm higher on the east side of the trench than on the west.

Again the relations in Trenches HA and HB are not clear. Units H:71B1–4 yielded flint and bones, but snail shells are not mentioned. Some snails are shown, however, on the section drawing. In any case, these units are not rich snail middens typical of Member T1 and thus are part of Member T2 or possibly even part of Stratum S2. The radiocarbon date from H:71B2–3 is equivocal; it seems too old for Stratum T, but its depth seems too shallow for Stratum S2.

No sediment sample was taken from this T1 midden.

Stratum S2

> Units H1B:166–172(west) and 159–172(east); H1A:186–204(east) and 190–204(west); H:60A2–10; (H:71B1–4?). This stratum is absent in FAS and its identification is questionable in subtrenches HA and HB.
> Dates: 12,540 (H1A:199); and probably 11,930 B.P. (H:71B2–3).

This stratum, like those below it, is dominated by a jumble of large limestone blocks, 1 m or more in length, from ceiling collapse, especially in FAS and H1B (Plate 5a, bottom). In H1A the rocks are smaller in general, suggesting that the rock falls were somewhat localized. In any case, they represent a different kind of rock fall from the catastrophic collapse that resulted in the opening of the big window in the middle of the cave in late or post-Neolithic time. Stratum S2 is quite variable in thickness, largely as a function of the localization of the large limestone blocks, and thus ranges from only 5 to nearly 100 cm thick.

The sedimentary matrix among the large blocks is (dark) reddish brown to very dusky red (5YR 3–4/3–4 to 10R 2/2) clay loam, and it varies from loose and/or dry to moist, sticky "terra rossa," enclosing conspicuous "white rocks." A moderate amount of bones—some well preserved—and flints are present, but few or no shells. Some excavation units had essentially no small rock fragments, whereas others enclosed "many small, round (waterworn?) pebbles" (H1A:192 in NB 519:89), which also had white surface alteration.

Sediment samples (FR 4–2 and 4–3) are dark yellowish brown (10YR 4/4, dry) silty or fine sandy loams with a few stones and bones, but almost no shells.

Stratum S1

> Units H1B:173–180; H1A:205–210; and FAS:208.
> Dates: none.

Another very rocky stratum with large roof-fall blocks in a clayey matrix. In this case, the stratum is recognized in both FAS and HH1. Its thickness ranges from 20 to ca. 60 cm in general, and even to as much as 110 cm in areas of very large blocks, e.g., SE corner of FAS (Figure 4.1).

This sediment is not much different from that of Stratum S2 except for the lack of "white rocks" (except at the very top—H1A:205 only) and the increase in small rock fragments in the clayey matrix. In FAS:208 the sediment is characterized by the excavators as "more rock than soil," and H1A:208 includes a "gravel lens." The rock fragments here tend to be more angular than in S2, and, in particular, unit H1B:177 is reported to contain "many small angular fragments and chips of limestone." Excavators called the colors reddish brown (5YR 4/4), red (2.5YR 4/6–8), and brown (7.5YR 4/4).

Sediment samples (FR 2–12, 4–4, 4–5) are light brown, reddish yellow, to pink (7.5YR 6/4, 7.5YR 6/6 and 7.5YR 7/4, respectively; dry) very stony clay loams, with very little cultural debris. The coarse sand grains are conspicuously rounded in FR 4–5. Note that there is a distinctive color change from 10YR hues above to 7.5YR hues in this stratum. Snail shells are present, and especially numerous in the sediment sample FR 2–12 from FAS:208, although none were reported by Whitney-Desautels. The equivalent stratum in Trench HH1 has a few snail shells, but not the large *Helix figulina* (Whitney-Desautels, forthcoming).

Stratum R

> Units FAS:209–217; H1B:181–212; H1A:212–220.

Dates: 21,480±350 (H1B:191–192) and 22,330±1270 b.p. (H1A:219).

This sediment is generally similar to Strata S1 and S2, but the excavators recognized certain differences. In Trench FAS the top of Stratum R (FAS:209) was called "a new deposit . . . totally different from the wet clay above the rockfall," i.e., S1. Here it is a yellowish red (5YR 4/6–8) clay loam with abundant rock fragments of all sizes and contains a number of animal bones, some flints, and dripstone fragments. Lower in FAS the excavators noted that the matrix seemed sandy, and they characterized FAS:212 as a yellowish sand, actually coded as 7.5YR 5/6 (strong brown), and it also included a "split sea pebble." Unit FAS:212 included well-sorted "beach sand?" along with bits of marine shell. Patches of reworked tephra from Stratum Q appear in units near the base of Stratum R in FAS, but were not recognized as such by the excavators at the time.

In Trench H1B a change was also noted at the top of Stratum R to "a sediment unlike anything dug yet." Sandy sediments were reported in several horizons within Stratum R, e.g., in H1B:196 a well-sorted sand with bits of marine shells. The lower units in Stratum R (e.g., H1B:210–212) include patches or particles of reworked tephra from Stratum Q.

Sediment samples (FR 2–13 and FR 4–6 to 4–10) are light yellowish red to reddish yellow (7.5YR 6/5 to 6/6, dry) stony clay loams enclosing some bone fragments in nearly every sample. Marine shells and microvertebrates were also noted in these samples.

Stratum Q

Units FAS:218–222, these units are mostly laterally adjacent, not superposed; and H1B:213.
Dates: none.

This unique stratum contains volcanic tephra that was deposited on a very uneven surface in and around large roof-fall blocks. The very light brownish gray (10YR 6/2, dry), fine-grained tephra is 5 to 9 cm thick where it is best preserved in FAS, with a very sharp lower contact and an upper contact that grades into the overlying reddish yellow, normal cave sediment (Plate 7b). In H1B:213 the tephra is more scattered and diffuse, probably owing to blowing and washing of the tephra shortly after its infall. In all cases, individual excavation units included not only the tephra but adjacent cave loam as well, largely because the excavators did not recognize the identity and importance of the tephra at first.

Sediment samples (FR 2–16, 2–18) are pure brownish gray (10YR 6/2, dry) tephra from FAS:221, and another sample (FR 4–11) is partly cemented tephra mixed with red loam from H1B:213.

Stratum P

Units FAS:224–227, and much of FAS:222 and 223, although crosscut; H1B:214–215.
Dates: none.

Essentially identical with Stratum R above the tephra. Yellowish red (5YR 4–5/6–8) clay loam with abundant gravel and rock fragments up to 25 cm across, in and around much larger limestone blocks. Thickness at least 1.5 to 2 m, as exposed in FAS; the bottom is below the groundwater table (i.e., sea level), which was encountered at a depth of 11.18 m below datum, or 1.21 below the official sea-level datum. Small yellowish brown particles, decayed particles of mammal bone or limestone, occur throughout (Plate 7b). Some flint and identifiable mammal bones are present in deepest exposures. In H1B much less of Stratum P was exposed, and, being higher above present sea level, it was hardpacked and cemented. Excavators reported that the sediment was sandy in H1B, and it included the same light brown particles as in FAS, but no finds. The bottom of H1B was nearly filled with large limestone blocks that could not be displaced, even with explosives.

Sediment samples (FR 2–19, 2–20, 4–12) were reddish yellow to strong brown (7.5YR 6/6 to 5/6, dry) clay loams with variable amounts of limestone fragments, all with black (Mn?) surface stains. Some large pale yellow blotches, up to 2 to 3 cm across, appeared to be thoroughly decayed limestone fragments. The sample FR 4–12 from H1B:215 was cemented.

SUMMARY

The lithostratigraphic sequence from Z to P is summarized in Table 4.2:

Table 4.2. Stratigraphic Summary

Stratum	Member	Thickness	Description
Z		ca. 200 cm (HH1) thinner in FA	Light grayish brown to dark yellowish brown loamy rock rubble with scattered ash and charcoal; very irregular lower contact. Mixed cultural remains.
Y3		ca. 100 cm (max.)	In A-FA-FF1 only; pale to dark yellowish brown sediment with hearths, ash and charcoal; large Neolithic pits.
Y2		ca. 200 cm (FA) ca. 100 cm (HH1)	Grayish to yellowish brown sediment with variable stoniness, including some 30 to 35 cm rocks in clusters in proximity of hearth complexes; lenses of *éboulis secs,* and occasional large blocks (80cm) from rock fall.
Y1		0–40 cm	Thin to discontinuous (in HH1) "rocky red" or "gritty" stony loam; most rocks less than 32 cm.
X2		35–80 cm	Moderately thick "gray clay" stratum relatively free of rock fragments; actually pale brown to light reddish brown silt loam; little cultural debris.
X1		40–60 cm (FA) thin in HH1	Conspicuous "rocky red" marker bed of sandy loam with abundant stones in 128 to 32 mm class; matrix yellowish brown to dark reddish brown; punky, white rocks, and rich in crushed snails and marine shell.
W	W3	50-60 cm	Dark (reddish) brown to yellowish brown silty loam midden rich in snails, ask, and charcoal; rather rocky, but less so than X1; appears transitional to X1.
	W2	60–70 cm (HH1) 160–175 cm (FA)	Thickest member of Stratum W; a complex of charcoal and ash lenses, hearths, and crushed snail shells in a dark reddish to yellowish brown loamy matrix; rock fragments particularly abundant in lower part, some with punky, white surfaces.
	W1	50–60 cm (HH1) 10–30 cm (FA)	Change in sediment type to "redder, shelly clay loam," actually dark brown to light brown moderately stony loam in FA samples; some snails and hearths but poorer in cultural debris than W2 or W3.

Stratum	Member	Thickness	Description
V		15–25 cm	Relatively thin, but conspicuous *éboulis secs* marker bed with sharp upper contact; yellowish brown to dark reddish brown, sparse matrix; 50–60% rocks between 128 and 32mm, many with thick white rinds; some snails and charcoal, but few artifacts.
U		40–50 cm	A "new deposit" of reddish to yellowish brown, compact clay loam with little cultural debris; fewer 128–32 mm rocks than V, but they are also weathered chalky white and rounded; all snail shells chalky.
T	T3	5–ca. 30 cm	In Trench HH1 only; very rich snail-shell midden with whole and crushed shells in a dark clay loam matrix; moderate amount of flint, bone and small rocks; all shells chalky.
	T2	5–25 cm	In Trench HH1 only; very loose midden of snails, charcoal, marine shells, animal and fish bones; fewer snails than in T3, but all are chalky; dusky red to yellowish brown matrix; stonier than T3.
	T1	10–30 cm	In Trench HH1 only; rich snail midden with many animal bones and a few large rocks; loose, dark reddish brown matrix; very irregular configuration because of underlying roof-fall blocks.
S2		5–100 cm	In Trench HH1 only; large limestone blocks from roof collapse with dark yellowish to reddish brown silty-sandy loam matrix ("terra rossa") with conspicuous white rocks; very few shells, but moderate amount of bones and flint; very few small stones.
S1		20–110 cm	Very rocky layer with large roof-fall blocks like S2, but no white rocks and more small rock fragments that tend to be more angular than in S2; clay loam matrix colors change from 10YR hues above to 7.5YR in S1 and below.
R		40–100 cm (FA) ca. 150 cm (HH1)	Generally similar to S1 and S2, but excavators recognized a "new" and "different" sediment—yellowish red clay loam with rock fragments of all sizes lying between and around large limestone blocks; abundant animal bones, some flints, marine shell, and dripstone fragments; some sandy horizons noted.
Q		5–9 cm (FA) diffuse (HH1)	Light pinkish gray, very fine-grained volcanic tephra, with a very sharp lower contact (in FA), but gradational upward into Stratum R; cemented in parts.
P		>150–200 cm	Very similar to Stratum R; reddish yellow clay loam with abundant gravel and small rocks up to 25 cm, deposited around large limestone blocks; small yellowish brown blotches of decayed limestone or bones; some flint and intact bones throughout, partly cemented in HH1; bottom not reached.

CHAPTER FIVE

Sedimentology

INTRODUCTION

This chapter describes the laboratory studies carried out on sediment samples from Franchthi Cave and their significance. The importance of such sediment studies lies in the information that they provide regarding the source and mode of deposition of the sediments and their postdepositional history, such as weathering, cementation, and perhaps erosion. Even more fundamentally, these studies provide objective, qualitative, and quantitative descriptions of the sedimentary strata that can be compared from layer to layer, from trench to trench, and from site to site. Laboratory analyses provide details on grain size, chemical and mineral components, weathering (roundness and porosity) of the stone fraction, and related parameters that cannot be evaluated by field inspection alone, especially in such polygenetic sediments as those that accumulate in an inhabited cave. For additional background, rationale, and procedures on cave sediment analysis, see Farrand (1975, 1985).

SAMPLING PROCEDURES

When I sampled in Franchthi Cave in 1976 (the final excavation season), only Trench HH1 was being excavated and in only the lowest levels of H1B. Trenches FAS and FAN were reasonably accessible, but the lowest meter of FAS was filled in. Trench H2 was badly overgrown with vegetation and partially collapsed; FF1 was slumped; and we did not clean the faces of Trench GG1, which had been standing exposed since 1968. Moreover, in 1976 it was already determined that much of FF1 and the upper two or three meters of GG1 had suffered significant modern disturbance and were not stratigraphically reliable. Thus, sampling was carried out in FAN, FAS, and H1B.

Samples were collected from a vertical face of each trench in a column about 20 to 25 cm wide. The columns are indicated on Plates 15 (H1B South), 14 (H1B West), and 7 (FA West) in Jacobsen and Farrand (1987) and on Figures 4.1 and 4.2 herein. Samples were not taken continuously down the column, but a sample was collected from each sedimentary layer that appeared physically different from those above and below. In the case of a relatively thick layer that appeared to be essentially uniform in character, more than one sample was collected in order not to miss some subtle variations. The samples were labeled with the column number (FR 1, FR 2, FR 3, or FR 4) and the position in that column. For example, sample FR 3-4 is the fourth sample from the top in column FR 3. A listing of all the samples along with a brief field description is given in Appendix A.

Each sample was excavated from the section face by digging 10 to 15 cm horizontally into that face. The vertical dimension of a sample was determined by the stratigraphy, so that no contrasting layers were mixed, to the extent possible. The dry weight of the samples ranged from 2010 to 10,400 g, and 75 samples averaged just over 5000 g each. In general, larger samples were taken from coarser layers in order to approximate a statistically valid sample of the largest particles present, and smaller samples were collected from fine-grained layers.

The stratigraphic thickness represented by an individual sample is always less than the total thickness of a lithostratigraphic unit and is usually thicker than an individual excavation unit. Recall that some excavation units were arbitrarily limited to only 5 or 10 cm in thickness in order not to miss conceivable changes in cultural, faunal, or botanical content, even though there was no apparent change in sediment type in the excavator's opinion. An idea of the sediment sampling density relative to excavation stratigraphy is given in this table:

Column FR 1: 32 samples from 166 units in FAN
Column FR 2: 20 samples from 40 units in FAS
Column FR 3: 15 samples from 50 units in H1B south face
Column FR 4: 12 samples from 58 units in H1B west face

for a total of 79 samples from 314 units, or an average of four excavation units per sediment sample.

ANALYTICAL METHODS

General

The procedures employed here are described and discussed in Farrand (1975). They are summarized briefly below.

Granulometry

Particle-size analysis began in the field. The bulk samples collected in the cave were dried in the sun, weighed, and sieved into two fractions—128 to 2.8 mm and <2.8 mm. (The 2.8-mm screen was the only one available in the field.) Any rock fragments >128 mm were excluded from the samples, but recorded. One hundred twenty-eight mm, or -7 phi, was chosen as the upper limit because it is a whole number on the sedimentologist's logarithmic phi scale, and is thus preferable to a limit of 100 mm commonly used in other cave sediment studies (Laville et al. 1980). From a practical viewpoint, inclusion of fragments larger than 128 mm would require impractically large samples in order to approach statistical validity.

In the laboratory the <2.8 mm fraction was first dry-sieved in order to obtain a representative sample of the fraction <0.125 mm destined for silt and clay analysis by the hydrometer method. Then, the fraction between 2.8 and 0.125 mm was wet-sieved. Rock fragments between 128 and 32 mm were measured by hand to determine their longest dimension.

The granulometric data are presented in histograms (Figure 5.1 and Appendix C) in order to emphasize the polymodal distribution of particle sizes, as discussed in Farrand (1985:31–32). Analysis of the individual histogram plots was aided by the technique of summing each one-phi bar for all 75 samples and plotting a grand average histogram that combines all the samples, which is shown in Figure 5.1. This technique suppresses the influence of the occasional sample with an unusually prominent mode and reinforces the modes that are common throughout the sample series. Also shown in Figure 5.1 is the frequency of occurrence of the common modes across all the samples. This analysis revealed the persistent occurrence of five size modes that peak between 128 and 32 (coarse rock fragments), 32 and 2 (small gravel), 2 and 0.063 (sand), 0.063 and 0.002 (silt), and <0.002 mm (clay) (Figure 5.1). Of these, the sand peak between 2 and 0.063 mm is the weakest, being most conspicuous in Strata W1 and W2; see sample FR 3-8 in Figure 5.1. Recall that Strata W1 and W2 are rich in

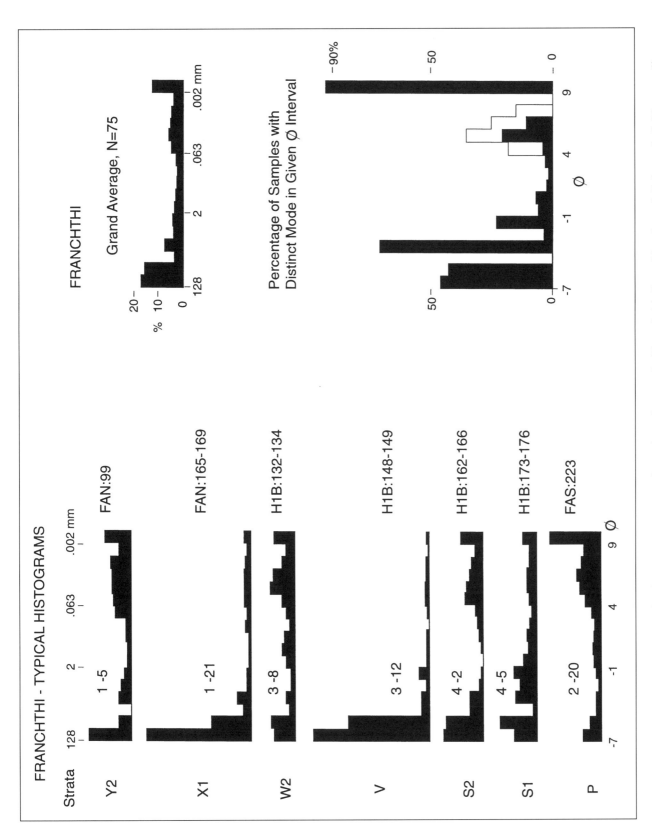

Figure 5.1. Typical granulometric histograms representing selected strata in Franchthi Cave. Numbers 1-5 through 2-20 are sediment sample numbers (see Appendix A). Histograms representing Strata X1 and V depict *éboulis secs*, while the others show prominent peaks of silt and clay. Also shown is a histogram of all 75 samples averaged together and a histogram showing the percentage of samples that include a peak in the five modes—coarse rock fragments, fine gravel, granules and sand, silt, and clay. The open histograms include samples in which the silt mode is present, but more diffuse, e.g., the sample from Stratum Y2 on the left.

snail-shell middens, and, in fact, much of the sand fraction in these samples is sand-sized snail-shell fragments. The five common modes are plotted in the global diagram (Figure 5.2) in terms of the weight percentage of the total sample that falls within the stated limits.

Roundness and Porosity

The physical condition of the rock fragments is particularly important in assessing the alteration or weathering of a given layer of sediment after deposition. With the exception of a few "rounded beach pebbles" observed by the excavators, the limestone fragments included in a given sample were presumably angular to subangular when deposited. For the most part, they had simply fallen from the roof or walls of the cave (Plate 6b). If they lay exposed on the cave floor for a prolonged period of time, they would be subject to solution weathering, which would attack first the angular corners and edges of the fragments, causing them to become increasingly rounded. Moreover, the solution process would attack the entire surface of the rock, producing a chalky white residue, as observed in numerous horizons in the sequence.

The degree of roundness was determined by classifying the fragments in a given sample as rounded, subrounded, subangular, or angular. On the rounded fragments, all the corners and edges were smoothed; on the angular fragments, all the corners and edges were still sharp. The intermediate classes were judged to have more rounded corners than angular ones (subrounded), or vice versa (subangular). See Farrand (1975) for further illustration of this technique. The results of this analysis are plotted in terms of a roundness index, calculated as follows:

$$\text{Index} = [(\% \text{ ang} \times 0) + (\% \text{ subang} \times 1) + (\% \text{ subrd} \times 2) + (\% \text{ round} \times 3)] / 3$$

Thus, an index of 0 indicates that all fragments in the sample are angular, and an index of 100 means all the fragments are rounded. The maximum value of roundness for the Franchthi samples is 68%, but most samples have an index of 50% or less (Figure 5.2).

Another result of solution weathering is the pervasive attack of water into the interior of a rock fragment, causing solution of fine interstitial grains and cement. The observed result is an increase in the porosity of the rock fragment. The bulk porosity of fragments was measured on air-dried rock fragments of a selected size, in this case 16 to 32 mm, that were immersed in water for 1 to 2 hours until they attained a constant weight. The weight increase relative to the initial dry weight yields the percentage of porosity, which varies from <1% in the least weathered fragments to 3.3% in more strongly weathered horizons (Figure 5.2). One outlier is a value of 9.5% from the bottommost sample, FR 2-20, which was collected very close to the groundwater level and presumably has been saturated ever since sea level rose to its present level several thousand years ago.

Calcium Carbonate, Organic Matter, and pH

The organic matter and calcium carbonate ($CaCO_3$) content of the fine fraction (<2 mm) were determined by loss-on-ignition (LOI) in a muffle furnace at 550° and 1000°C, respectively. Samples were run at least twice to obtain consistent readings.

In addition, the fine fraction was subjected to a solution of 50% hydrochloric acid until reaction ceased. The acid-soluble portion of the samples was consistently higher than the LOI $CaCO_3$ by ca. 10 to 25%, but the values paralleled the LOI values, as can be seen in Figure 5.2. Presumably the acid attacked not only the finely divided $CaCO_3$, but also less soluble components such as charcoal and shell fragments, as well as apatite from fine splinters of bone and collophane cement.

The LOI CaCO$_3$ values range from ca. 20 to ca. 60% of the initial weight of the fine fraction, and they show systematic broad peaks and troughs that correlate well with other parameters such as pH, roundness index, chalky snails, and to some extent with porosity in Figure 5.2.

The acidity (pH) of the samples was determined on an aqueous slurry of the fine fraction measured by an electronic pH meter. The pH values range from ca. 7.1 to 8.2 and show a good correlation with CaCO$_3$ values (Figure 5.2) as expected, i.e., high (alkaline) pH values match peaks of CaCO$_3$ and vice versa.

CULTURAL CONTENT OF THE SAMPLES

Discrete residues of cultural activity that accompanied sediment accumulation were extracted from each sample. These residues included sherds, flint debitage, charcoal fragments, bones, and marine-mollusk and land-snail shells. The abundance of these materials is plotted as a weight percentage of the total sample, and the values range from <1 to ca. 7.5%. Very few samples were completely devoid of cultural debris.

An important measure of cultural influence is the abundance of land-snail shells in a given sample on the assumption that the snails were a food resource. The data presented here (Figure 5.2) are visual estimates of the percentage of snail-shell fragments retained on the #18 (or 1-mm) sieve during wet sieving, thus fragments measuring between 1 and 2 mm. In snail-midden layers (Strata W2 and T3) the percentage is as high as 62.5 to 87.5%. The plotted values show that snail-shell fragments are present in nearly all samples, but they are rare near the bottom of the sequence in Strata P, Q, R, and S1. This plot is in agreement with the data of Whitney-Desautels (forthcoming), who shows that the large *Helix figulina* are not found in these lower strata, although some smaller snails do occur there.

Other information derived from the study of land snails in the cave sediments is very informative in interpreting the sedimentary sequence. In Whitney-Desautels' (forthcoming) study, she found that some snail shells had retained their shiny surfaces, while others were chalky or crusted with CaCO$_3$, and still others occurred only as external casts. In some layers only shiny shells occurred; in others only chalky shells, or there was a mix of shiny and chalky shells. This information is plotted stratigraphically on the right side of Figure 5.2, which shows an extended interval from Stratum S1 through Stratum U without any shiny shells. There were too few shells in Strata Y3 through X2 to use reliably for this kind of evaluation.

Snail casts are somewhat problematical. These are casts formed in a cavity (mold) filled with sediment after a shell had weathered away. They occur frequently in the lower strata and again in the uppermost strata and represent many of the smaller snail species, but not the large *H. figulina*, which is uncommon or absent in strata where casts occur. Whitney-Desautels (forthcoming) considered the possibility that the casts were intrinsic to the lower, clay-rich Palaeolithic strata and that their recurrence in the upper levels of the sequence was an indication of reworking of Palaeolithic strata somewhere else in the cave. However, I consider this unlikely for two reasons. First, the casts in the upper strata include *Rumina decollata*, whose stratigraphic occurrence as intact shells seems to be restricted to Holocene levels. Secondly, any reworking would have to have been very extensive to mix sediments that occur six or more meters below the surface into the topmost one or two meters. There is no independent indication of such deep disturbance, although our excavation area covered only about 10% of the cave floor. On one hand, my colleagues report that the units above FAS:59 and FAN:59 appear to be almost "pure" Palaeolithic in terms of lithics and fauna, and some of these presumably Neolithic units also contain Mesolithic components; see also Vitelli (1999). On the other

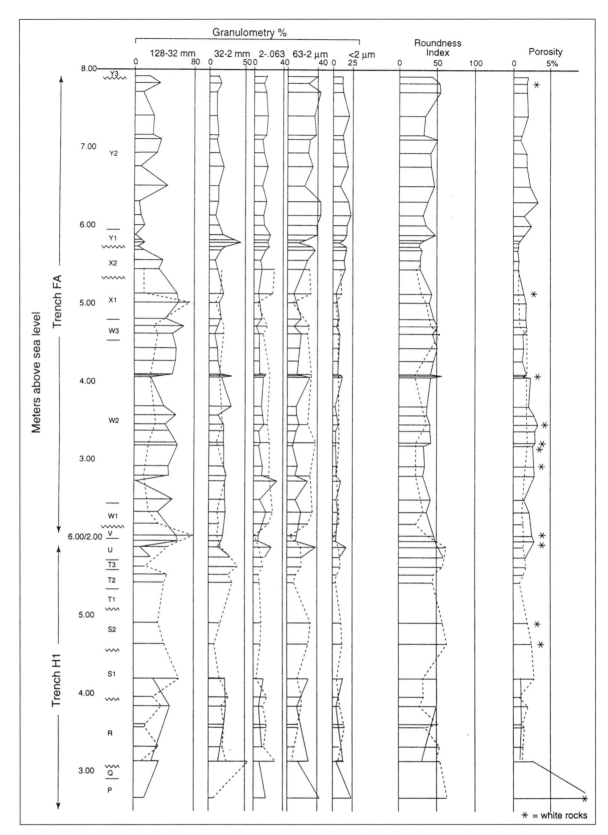

Figure 5.2. Global diagram of sedimentological parameters used in this study. Sample locations are plotted against the vertical scale in meters above sea level, with the upper part being based on Trench FA and the lower part on Trench H1B because of the major temporal hiatus in Trench FA (see text). In each part, the samples from the other trench are plotted in their stratigraphically equivalent positions, not according to their altitude (see Table 4.2). The zigzag lines connecting the values in each column are drawn as solid lines for FA samples and as dashed lines for H1B samples. In the case of "CaCO$_3$," the solid and dashed lines are loss-on-ignition (LOI) values, and the dotted line gives the acid-soluble values for both FA and H1B. The five granulometric columns (at left) show the percentage by weight of each prominent mode, as discussed in this chapter and illustrated in Figure 5.1.

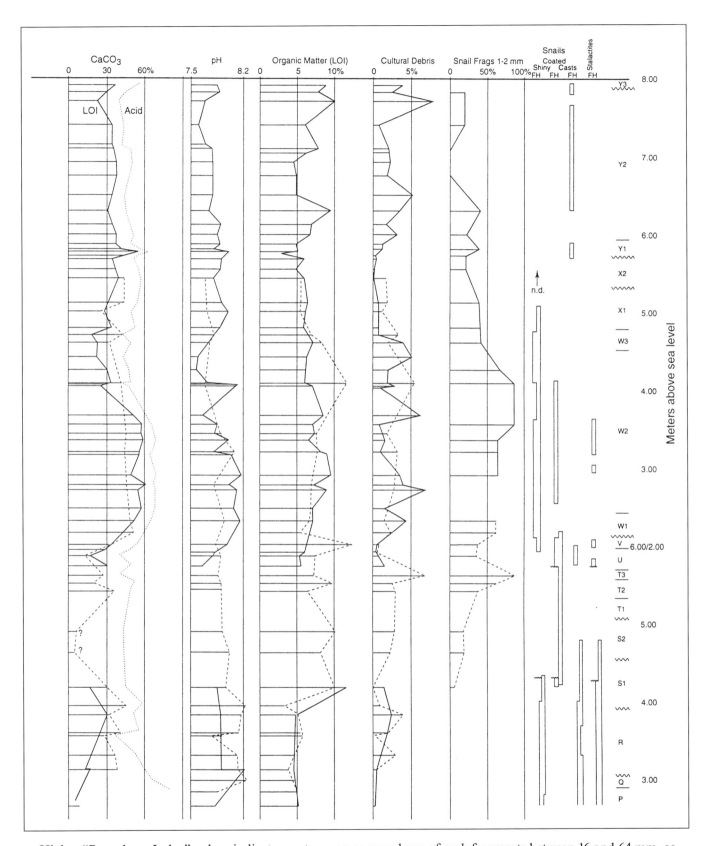

Higher "Roundness Index" values indicate greater average roundness of rock fragments between 16 and 64 mm, as explained in the text. Bulk "Porosity" was measured on fragments between 16 and 32 mm. The "Cultural Debris" is the weight percentage of sherds, lithic debitage, charcoal, bones, and shells in the >2 mm fraction of each sample. "Shell Fragments" is a visual estimate of the percentage of land-snail-shell fragments relative to rock and mineral grains of the same size retained on the 1-mm sieve during wet sieving. The condition of snail shells and the occurrence of stalactite fragments recovered the wet-sieving operations on the site (N. Whitney-Desautels, forthcoming) are indicated as "shiny" (or normal), "coated" (or chalky), or as "casts" of shells that had otherwise disappeared; these are presence/absence plots for Trenches FA ("F") and H1B ("H"). See text for additional descriptions of the analytical results; they are also tabulated in Appendix B.

hand, Whitney-Desautels' (forthcoming) data come only from units below FAS:59 and FAN:59; she did not sample the higher units. To my knowledge, no reworked Palaeolithic materials were reported from the Neolithic units below FAS:59 and FAN:59. This casts doubt on the reworking hypothesis to explain the snail casts. An alternative explanation might relate to conditions appropriate for fossilization or preservation that were similar in the upper levels of Trench FA to those in the lowermost strata. Moreover, it must be noted that Whitney-Desautels did not sample levels in Trench HH1 equivalent to those with casts in the upper levels of FA. Thus, there is no independent confirmation of their occurrence elsewhere than in Trench FA.

Finally, in the process of sorting water-sieved residues for snails, a number of stalactite fragments were recovered. These were pieces of small, "soda-straw" stalactites, some of which can be observed at present on the cave ceiling, although they were not actively dripping in the summer of 1976; see Figure 5.2 and Plate 3b. (Their concentric structure looked superficially like coiled snail shells to the young Greek girls sorting the residue; thus they were put into the snail collections.) Stalactite fragments were also found in my sediment samples, and some were noted by excavators during excavation, especially in the lower part of the sequence. My sediment samples, however, were much smaller than the volume of sediment that went to the water-sieving operation, so that the probability of encountering these fragments was reduced accordingly. Nevertheless, all three sources of information (sediment samples, excavators' notes, and the "snail" collections) concur that stalactite fragments are common in the middle of Stratum W2 and from Stratum U downward, particularly in Strata S1, R, and P.

I should point out, however, that the collection of these data was not done in a systematic fashion, except for the sediment samples. In the first place, serendipity was involved in finding the fragments in the snail collections, and, secondly, the excavators were not systematic in recording such materials in the trenches. The process that contributed the stalactite fragments to the accumulating sediment is not clear, but it was likely a natural, rather than anthropogenic, process given the abundance of fragments in the lower strata, where evidence of human presence is strongly reduced.

PRESENTATION OF DATA

The analytical data are tabulated in Appendix B and are assembled stratigraphically in a global diagram (Figure 5.2) in which the vertical axis is absolute height in meters above sea level. The position of each sample is shown by a horizontal bar to emphasize the fact that the samples are neither regularly spaced nor continuous. The data for the two trenches are superposed in stratigraphically equivalent positions. Data points from Trench FA are connected with a solid line, and those from HH1 with a dashed line. However, in order to arrange the data appropriately, the vertical scale is divided into two sections, because some strata (T3, T2, T1, and S2) do not occur in Trench FA, and levels equivalent to the upper part of the FA sample sequence were not sampled in HH1. Thus, the upper part of the y-axis is scaled to sample heights in Trench FA, i.e., down through Stratum W1, and the samples from HH1 are placed in their correlative positions. The lower part of the y-axis is scaled to heights in Trench HH1, from Stratum V downward, and the FA samples are adjusted to correlative positions. Moreover, the heights of the samples have been normalized (or projected) to a common corner of a trench complex. This was necessary because of the dip of the stratification, as seen in Figures 4.1 and 4.2. The height of an excavation unit in one trench may be as much as 25 cm higher at one extreme than the extreme corner of the contiguous unit in the adjacent trench. Thus, for the Trench FA complex, heights are normalized to the common corner on the west face of the trench, i.e., where the NW corner of FAS meets the SW corner of FAN. In the HH1 complex, heights are normalized to the SW corner of Trench H1B.

The snail and stalactite data from Whitney-Desautels (forthcoming) are plotted as vertical bars spanning the stratigraphic interval in which the indicated conditions were found. These are presence/absence data. Data from Trench FA are on the left of each paired column, and HH1 data are on the right.

Concerning correlation of the samples between Trenches FA and HH1, I wish to emphasize the point that was made in the introduction to Chapter 4. The correlation of lithostratigraphic units among the various trenches (Table 4.1 and Figure 6.1) was based on marker beds, macroscopic similarities, and radiocarbon dates, not on the analytical data in Figure 5.2. This being the case, it is interesting now to compare the analytical results for correlative strata in Trenches FA and HH1; in other words, how do the solid lines compare to the dashed lines in Figure 5.2? This will be a measure of the uniformity of sedimentation throughout the part of the cave represented by these two trenches. This comparison can be made only for samples from Stratum X2 downward, since the Y strata were not sampled in Trench HH1.

One can see that the agreement, in general, is quite good, in particular for roundness, porosity, pH, and organic matter. In the granulometry columns, the clay (<2 mm) values match quite well, as do the sand (2–0.063 mm) percentages. (Note that the density of sampling points is greater in Trench FA than in HH1 in Strata W2 and W3, which means that there is less detail for HH1 in this part of the sequence.) The 128–32 mm column indicates that Trench HH1 is less rocky in the Mesolithic levels than is the case in FA, although the values for the two marker beds (*éboulis secs* X1 and V) are very similar in both trenches. This close agreement from trench to trench is an argument in support of using a limited number of sampling columns to represent all the excavated areas in the cave. On the other hand, the observed differences between the sampled trenches may be attributed to real differences in sedimentary microenvironments across the cave floor, coupled with different functional areas for human activities. The latter conclusion is supported by trench-to-trench variations in the cultural debris column, and to some extent in the organic matter values.

INTERPRETATION OF THE SEDIMENTOLOGICAL DATA

From the descriptive data discussed above and presented in Figure 5.2, one can move towards an interpretation of the origin of the sediments and their postdepositional history. The next step will be an attempt to reconstruct the palaeoenvironment under which the inhabitants of Franchthi Cave were living.

Sediment Origins

As I have explained elsewhere (Farrand 1975, 1985), sediments in caves inhabited by prehistoric peoples are polygenetic in origin. Natural, or geogenic, processes may produce sediment intrinsic to the cave. Bedrock of the walls and ceiling may be broken down by solution, freeze-thaw, earthquakes, etc. and accumulate on the cave floor. In some caves, running water from the cave interior may transport clayey residues into habitation areas. Dripstone (stalactites, stalagmites, travertine) may be deposited by the evaporation of water dripping through the cave ceiling. Extrinsic materials can also be introduced into the cave by running water, by mass movements of slope debris, by wind transport, and by seepage down through fissures (or "windows," as in the case of Franchthi).

Anthropogenic activities also are responsible for bringing sedimentary components into a cave, intentionally or otherwise. Food, clothing, wood for fires or for structures, lithic materials for making tools, and ceramics are brought in during routine activities. Stone structures may also be built from rocks collected outside the cave. Moreover, the inhabitants may dig pits into the sediments on the cave floor or move sediment around to level the floor. Another very important component is mud, sand, clay, soil, and even vegetation that may be introduced unintentionally, being attached to peoples' feet, skin, or clothing. Any of these anthropogenic components may be mixed with the accumulating geogenic sediment.

In a cave that is open to the exterior, almost any combination of the materials mentioned above may be accumulating simultaneously. One hopes that a detailed sedimentological analysis will reveal the several components and their relative importance in any given layer of sediment. In some cases, the relative importance of a certain component may be inferred from another component; for example, the quantity of exotic sand grains may well be correlated with the abundance of food debris in a given layer, but in the absence of indications of human presence, such exotic sand must be explained by some geogenic processes—wind, running water, etc. Similarly, a high value of $CaCO_3$ in the fine fraction may be the result of the abundance of crushed snail shells in that stratum and, thus, at least in part, is anthropogenic.

Sources of Franchthi Sediments

In the case of Franchthi Cave, one can see the relative importance of the original sedimentary components from inspection of Figure 5.2, namely in the columns for granulometry (five columns), organic matter, cultural debris, and snail fragments. The granulometry is largely geogenic in origin, but not entirely because, as discussed above and below, a large proportion of the sand-sized fraction is in fact snail-shell fragments of anthropogenic origin (see also below). The other three columns definitely reflect cultural input, and one can see that the peaks and troughs of all three components parallel each other to a very great extent with consistently high values in the strata with many snail middens—W1, W2, W3, T2, and T3. Independent support for a high level of anthropogenic input during the time of Stratum W comes from the palaeobotanical remains, the quantity of which per unit time is far greater than at any other period represented in the cave (Hansen 1991:157–164). Hansen's zones III, IV, and IV/V, which correspond to Member W2, have 1300 to 6500 g carbon per 100 years, whereas the preceding and following zones have <100 g per 100 years. In addition, the excavators noted the common occurrence of "polished sea pebbles" in these strata, presumably indicating rock collecting by the inhabitants during their visits to the beach.

The granulometric curves indicate that Strata V through X1, corresponding approximately to the time of Mesolithic occupations, have consistently large components of 128–32 mm rock fragments, especially in Trench FA (Plate 8d). Those same levels tend to have relatively large amounts of fine gravel-sized fragments (32–2 mm), too, and the smallest amounts of clay-sized sediment. Furthermore, the rate of sedimentation in these levels is the highest of the Franchthi sequence, in fact very high—perhaps as much as 2.5 cm per year (see Chapter 6). These stony Mesolithic strata stand out distinctly from both the overlying Neolithic and basal Upper Palaeolithic levels, which are dominated by a loamy to clay loam matrix, as indicated by the 63–2 and <2 μm columns in Figure 5.2.

The 128–32 mm rock fragments are almost entirely of limestone derived from the walls or ceiling of the cave. The cave bedrock is highly fractured as a result of tectonic deformation of this area of Greece. Recall that the dip of the bedrock strata in and around the cave is vertical or nearly so, indicating serious structural deformation in the geological past that subjected the bedrock to much internal stress and strain. A good view of the nature of the fragmentation of the bedrock can be seen at the far inner end of the cave (Plate 6b). Here the bedrock, although still standing as a solid wall, is fissured into irregular polygons ready to be detached by a slight amount of solution along the fissures or other disturbance, such as seismic tremors. In the photo the fragments are roughly between 20 or 30 to 150 mm across, thus in the same range as the 128–32 mm size mode in the sediment samples.

The exact mechanism for releasing the bedrock fragments is not known. Freeze and thaw is very unlikely in this climatic area. Ground temperatures of -5°C or below are required for effective freeze-thaw destruction of most rock types (Guillien and Lautridou 1970), and even air temperatures that low must be extremely rare in this part of Greece at present (van Andel and Sutton 1987:5–8). Given that the greatest input of bedrock fragments into the sediments occurred during the Holocene Epoch, one can safely assume that the climatic regime at that time was essentially that of today. Even during the time of the last glacial maximum, around 20,000 to 15,000 B.P., a temperature depression of at least 15°C below today's January monthly mean would have been required to induce significant freeze-thaw action, and that is significantly greater than the temperature depression for this region estimated by other palaeoclimatic indicators (e.g., CLIMAP project members 1976).

Some Franchthi strata are conspicuous because of their extreme abundance of rock fragments and minimal amount of fine matrix, thus resembling *éboulis secs* recognized in many French caves and rockshelters (Farrand 1975; Laville et al. 1980). *Éboulis secs* are open-work deposits of angular rock fragments, perhaps initially without any matrix at all. Such deposits are likely to have formed rapidly, so that there was little time for further breakdown of the fragments into finer components or for simultaneous input of fines from another source. *Éboulis secs* are commonly associated with freeze-thaw processes in cold climates, such as ice-age France, but they are known also from warm climates where freeze-thaw activity is very unlikely. I have observed them in Holocene rockshelter deposits at low altitudes in the southern Dead Sea rift and in southwestern Texas.

In Franchthi Cave two marker horizons are categorized as *éboulis secs,* Strata V and X1. They can be traced nearly continuously around the four walls of Trenches FA, FF1, and HH1 (see Figures 4.1 and 4.2; Plate 6a). Stratum V is probably represented also by units G1:66 and 67 at the bottom of Trench GG1, but Stratum X1 appears to be missing in that trench. On the granulometry plots in Figure 5.2, Strata V and X1 have the highest values of 128–32 mm fragments and essentially no clay- or silt-sized component. They contain almost no cultural debris other than a few snail-shell fragments, which might have filtered down from overlying sediments. Moreover, both of these strata appear to be succeeded immediately by an unconformity—a sedimentary hiatus—of perhaps 600 to 650 years (see Table 6.2), an interpretation supported by the radiocarbon chronology, by the sharp upper contacts of those strata, and by the presence of rock fragments with conspicuous chalky surface alteration. Perhaps coincidentally, each of these rocky layers occurs at the top of a major cultural stage. Stratum V is at the top of the Upper Palaeolithic sequence and X1 caps the Mesolithic occupations.

The origin of these *éboulis secs* layers is not certain, but their continuity within a given trench and from trench to trench suggest that they are geogenic, not anthropogenic. As stated above, they appear to have been deposited rapidly and without matrix. This is suggestive of earthquake activity, well known in this area, which might have detached many fragments from the walls and ceiling that were already loosened and ready to fall. On the other hand, the tremors must not have been strong enough to bring down major portions of the cave ceiling like the large blocks that fell in post-Neolithic time. Speculatively, such a geological catastrophe might have imprinted the memory of the local people so that they avoided the cave for a number of generations, or forever. Subsequent occupation of the cave was by a group with another cultural tradition, for example, as recorded by the Neolithic settlers who came into the cave after the time of Stratum X1.

Decalcified Granulometry

Another aspect of the granulometry concerns the composition of the fine fraction. It has already been mentioned that many of the particles in the coarse sand sizes (1 to 2 mm) are snail-shell fragments, as plotted in the snail fragments column in Figure 5.2. Shell fragments also occur in finer sand fractions, along with bone fragments, microfauna, charcoal fragments, and other finely divided remnants of cultural activities. Thus, it is interesting to consider the particle-size distribution after treating the fine fraction with hydrochloric acid, which dissolves the snail shells and bones as well as any carbonate or phosphate cement that may be binding some of the finer grains. The results are presented in Figure 5.3 by comparing histograms of the fine fraction before and after acid treatment. Only every other sample in the FR 1 and FR 2 samples series is depicted here. The medium to coarse sand sizes (2 to 0.125 mm) have been lumped in the open bar at the right of each histogram. The <0.125 mm distribution was determined by hydrometer analysis.

The first difference between nondecalcified and decalcified samples is the strong decrease, by a factor of two or three, in the medium and coarse sand-sized component in all samples. This is not surprising because much of that fraction is shell fragments, as already stated. However, the decrease in this fraction is not proportional to the percentage of shell fragments observed in the 1 to 2 mm nondecalcified fraction (cf. fragments column in Figure 5.2). The medium- to coarse-sand fraction in samples with <10% shell decreased proportionally as much as in samples with 37.5 to 50% shell fragments. Clearly other components were dissolved as well. The remaining, non-acid-soluble, material in this size range is composed largely of sand grains of various rock types that occur outside the cave, including some fine flint debitage in some samples.

The second consistent change is in the silt-sized range. In nondecalcified samples, the silt mode is usually clearly defined, but after acid treatment there tends to be relatively more silt, and it is broadly distributed across the silt range without a clear peak in many samples. The prominence of the silt fraction is probably a function of wind transport of dust from outside the cave. A long-traveled silt component has been suggested in other Mediterranean cave settings (Farrand 1979:374–376) and in terra rossa soils in Greece (van Andel 1998b:380). However, under the microscope, the Franchthi silt grains appear rather angular or subangular, suggesting that they have not traveled great distances. Some or most of this silt may have been blown from the exposed coastal plain in front of the cave at times of lower sea level. Also some silt may have been introduced along with terra rossa that washed into the cave from above.

Finally, the clay-sized material tends to decrease in relative abundance in most acid-treated samples, and almost disappears in several samples—FR 2-0, 2-2, 2-4, and 2-8. In only two samples the clay-sized material increases—FR 1-30 and 1-32. In the cases where the clay-sized material almost disappears, the original sample was strongly dominated by shell fragments, and in some of these samples there also were white, weathered limestone fragments, even in the sand sizes. I suggest that the clay-sized material in the nondecalcified samples was mostly very finely divided shell debris or the chalky dust from the surfaces of the white rocks and sand grains that dissolved in acid. It is interesting that a very strong peak appears in the fine silt range, between 7 and 8 phi (0.008 to 0.004 mm), in four of the decalcified samples in which the clay peak is reduced to near zero; see also the previous paragraph. In three of these four samples, another sharp peak appears in the finest sand range, between 3 and 4 phi (0.125 and 0.063 mm). This latter peak appears clearly also in 15 of the 25 samples compared in Figure

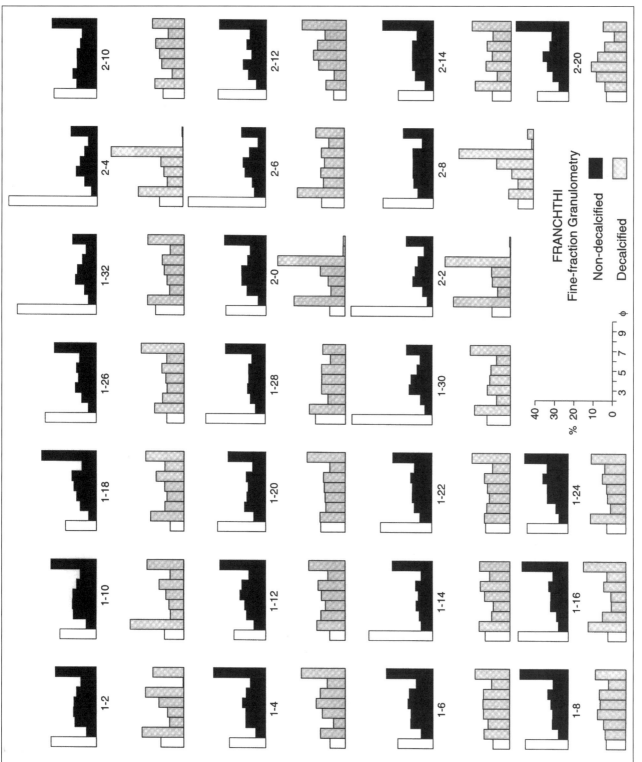

Figure 5.3. Histograms of non-decalcified and decalcified aliquots from selected samples from Trench FA. The histograms above and below the sample number represent the same sample. The open bar at left in each histogram includes the totality of the sand fraction of that sample. In general, it can be seen that the decalcified histograms show a more uniform distribution than do the non-decalcified ones, but significant shifts between the silt and clay fractions occur in many samples. See text for further explanation.

5.3. Presumably it originates upon the breakdown of the coarser sand-sized material, perhaps grain aggregates cemented with $CaCO_3$.

In both the nondecalcified and decalcified particle distributions (Figure 5.3), there is a consistent abundance of silt- and clay-sized material in the upper part of the stratigraphic sequence, namely Strata Y3, Y2, Y1, and X2. This loamy sediment coincides with the Neolithic occupations of the cave, during which the intensity of anthropogenic input appears to be less than in the preceding Mesolithic, and thus the origin of the loam is likely to have been due to a natural process. I speculate that one or both of the windows in the cave ceiling may have already begun to open, allowing the introduction of reworked soil from above the cave to enter freely and mingle with other sediment on the cave floor. There is no direct evidence for this scenario; it is presumably hidden below the huge, undated pile of breakdown in the center of the cave. However, it is very reasonable that the breakdown of the ceiling took place in stages, even though the final collapse was undoubtedly a major catastrophe. One supporting argument may be the absence of stalactites in the Neolithic sediments, which suggests that the cave was more wide open and thus drier during the accumulation of Strata X2 through Y3 than during the Mesolithic and Upper Palaeolithic occupations.

POSTDEPOSITIONAL MODIFICATIONS

Even as the original sediment is accumulating on the cave floor, it may be undergoing modifications, and the longer a layer of sediment lies exposed on the cave floor without being covered by younger sediments, the more it will be subject to postdepositional change. Some of these changes may be physical displacements of the sediment through human agency—digging pits, burials, leveling the floor, building walls, etc.—or through natural erosion by running water, or even through animal activities, such as hibernation, in some caves. Other modifications fall under the heading of weathering. In these cases it is the chemical action of water that dissolves or attacks certain mineral constituents. In limestone caves, such as Franchthi, the ambient water is charged with atmospheric carbon dioxide (forming carbonic acid) that attacks the $CaCO_3$ of the limestone, as well as of mollusk shells and animal bones. Corrosion or even complete dissolution of these materials may occur. This corrosion will cause rounding of sharp corners and edges of angular fragments and will commonly leave a white, chalky coating on the rocks and shells. Further, the chemical attack on bone may break down the bone mineral fluorapatite, $Ca_5F(PO_4)_3$, into soluble products, which in turn may be redeposited as amorphous collophane, filling vugs or cementing the sediment (Goldberg and Nathan 1975). In addition, in sediments with a large amount of organic matter (middens, other human refuse, or bat guano), humic acids may form and attack the mineral constituents.

Weathering of cave-floor sediment requires a certain amount of time and takes place largely at the sediment/air interface. As surface water attacks carbonate minerals, it gradually becomes saturated with carbonate ions and is no longer an effective reagent. If sedimentation is more or less continuous, there will be little opportunity for solution weathering to take place. Thus, a cessation of sedimentation, or at least a sharp decrease in sedimentation rate, is required for significant weathering to occur. Moreover, since cultural activities are responsible for a considerable amount of sedimentation, as we have seen above, one implication of a weathered horizon is the absence of intense human occupation at that time.

In the analysis of Franchthi sediments, the variations in roundness, porosity, $CaCO_3$, pH, and the conditions of snail shells are critical in the recognition of weathered intervals. The occurrence of "white rocks" noted by the excavators and in my samples is another, nonmetric parameter of importance; major white-rock horizons are marked with an asterisk in the porosity column of Figure 5.2. As discussed above, a corollary of the presence of weathering phenomena is cessation of sedimentation, and this will be examined in Chapter 6 in connection with the radiocarbon chronology.

The roundness index and porosity values (Figure 5.2) tend to covary, if weathering is involved. The attack of acid-charged water will round the edges of angular stones and penetrate to the interior of the rock fragments dissolving interstitial fine-grained minerals and cement. The porosity of the unweathered Franchthi limestone is <1%, but in some weathered horizons, porosity has increased commonly three-fold to ca. 2.5 to 3%. Note that roundness and porosity are both relatively high in the Holocene strata, Y1, Y2, and Y3, and again throughout the Mesolithic strata, W1, W2, W3, and X1, where white rocks are common, but a zone of relatively low values intervenes in Stratum X2, the "gray clay" of the Initial Neolithic phase (FCP 0/1, Vitelli 1993:38–39). The most pronounced and consistent rounding of rock fragments is found in Strata U and V, once again accompanied by relatively high porosities and chalky, white rocks.

Turning to the $CaCO_3$ and pH columns, which parallel each other very closely as expected, one finds that the $CaCO_3$ values vary by a factor of two or more—from ca. 60 to <30%. The strongest reductions occur in the upper part of the Mesolithic sequence, Strata W2 (upper), W3, and X1, and in Strata U and V at the top of the Palaeolithic sequence. These reductions match high roundness and porosity values mentioned in the previous paragraph.

The $CaCO_3$ and pH values are at odds, however, with the roundness and porosity figures in the lower part of the Mesolithic levels. The $CaCO_3$ values in W1 and the lower half of W2 are the highest of the entire sequence and are readily explained by the great abundance of snail-shell fragments, as well as by the rapid rate of sedimentation of those strata. Note also that chalky snail shells are found in these same levels in Trench FA, which reinforce the indications of weathering given by the roundness index and porosity values. Although snail shells are also abundant in Stratum W3 and the upper part of W2, the strong reduction of $CaCO_3$ in those layers is most likely to be attributed to weathering progressing downward from the unconformity at the top of Stratum X1.

The weathering indicators in Strata V and U are the strongest of the cave sequence. As noted above, the roundness index is quite high and $CaCO_3$ is consistently low. Moreover, the snail shells are nearly all chalky, and shell casts occur, too. A close second is the weathering in Stratum S2, with roundness index and porosity values nearly as high as those in U and V, and only chalky snails, along with white rocks. The $CaCO_3$ values are somewhat problematical, however. The $CaCO_3$ LOI numbers are suspiciously very low, only 5 to 6%, while the acid-soluble "carbonate" is relatively high, and pH readings appear to reinforce the acid-soluble data. Both of these weathered horizons, U-V and S2, underlie unconformities of ca. 500 years duration each.

THE BASAL BEDS

The lowest part of the Franchthi sequence appears significantly different. In the first place, there are many, very large limestone blocks scattered throughout these layers (Figures 4.1, 4.2). Secondly, in Stratum S1 and below, the sediment color changes from 10YR to 7.5YR hues, thus becoming somewhat redder. This sediment has a moderately high percentage of 128 to 32 mm rock fragments. A "gravel lens" was noted by excavators in Unit H1B:208, and FAS:208 was said to contain "more rocks than soil." In these strata the matrix was noticeably clayey and sticky to the excavators. The decalcified fine-fraction histograms for FR 2-12, 2-14, and 2-20 (Figure 5.3) show a consistently high, <0.002 mm clay mode and a better developed silt mode than most of the other decalcified samples. In fact, there are similarities with the sediments of Stratum Y2 near the top of the stratigraphic sequence, which were interpreted (above) to be reworked soil introduced into the cave from above.

Although the roundness index is relatively high, porosity is low in Strata S1 and R, but unusually high in FR 2-20 (Stratum R), the bottommost sample near the water table. CaCO₃ values are in the middle of the range of overlying samples, and pH values are average (in FA) to relatively high (in HH1). Shiny snail shells appear with depth in S1 and continue to the bottom of the exposed section, and there are no more chalky snails below the middle of Stratum S1. (Note that there are very few snails at all in these strata [Whitney-Desautels, forthcoming].) Casts of snail shells and stalactite fragments are persistent from the top of S1 to the base of the section. Although not abundant, evidence of human occupation of the cave is found throughout this lowest part of the stratigraphic sequence in the form of large mammal bones, particularly equids, chert artifacts, a few marine mollusks, and some charcoal.

This basal sequence is separated in time from the overlying beds by a major unconformity, spanning at least 6400 years, as discussed in detail in Chapter 6 and in Farrand (1993). Stratum S1 must be older than 14,710 Cal B.P., and is probably older than 17,360 Cal B.P. Stratum R has two dates of ca. 26,000 to 27,000 Cal B.P., and Stratum P is older than the tephra of Stratum Q, which is in the 33,000 to 40,000 Cal B.P. range. Thus the basal two meters span about 20,000 years, at least, which is a duration about twice as long as all of the overlying 8 to 10 m, although it was interrupted by significant hiatuses (see Chapter 6).

Therefore, the basal beds present a sequence that is strongly disjunct in time from the overlying strata, and they have a sedimentological character of their own. Recall, moreover, that these beds are not really the base of the sedimentary fill in Franchthi Cave. Electric resistivity measurements (Chapter 3) indicated at least another five or more meters of fill below the bottom of Trench FA, which is now near sea level.

SUMMARY

Laboratory analyses of 75 samples taken from Trenches FA and H1B shed important light on the origin and postdepositional modification of the Franchthi sediments (Figure 5.2). Granulometric analysis reveals (a) a predominantly loamy Holocene section, (b) Mesolithic levels rich in limestone fragments, poor in silt and clay, and with abundant anthropogenic input, (c) a somewhat rocky final Upper Palaeolithic sequence with prominent snail-shell middens, and (d) a basal sequence of sticky clay loams that enclose large limestone blocks and a thin bed of volcanic tephra. It is suggested that the loamy Holocene sediments of Strata Y1, Y2, and Y3 were derived from surface soils that washed into the cave through windows in the roof that were beginning to open in early Holocene time. Two marker beds, Strata V and X1, composed of *éboulis secs,* cap the Palaeolithic and Mesolithic cultural phases, respectively. They appear to mark intervals of temporary abandonment of the cave. The basal beds appear to belong to a different sedimentary regime that prevailed at a time when the cave was more closed than now, as indicated in part by the abundance of stalactite fragments in the basal sediments.

Effects of weathering are indicated by the roundness index and porosity of limestone fragments, reduction of CaCO₃ and pH of the fine fraction, the presence of chalky white rocks, and the condition of the enclosed snail shells. The results point to three distinct weathering intervals—at the top of the Mesolithic sequence in Strata X1, W3, and the upper part of W2; at the top of the Upper Palaeolithic sequence—Strata V and U; and in Stratum S2. Each of these intervals marks an unconformity of 200 to 500 years, as seen in Chapter 6, during which times the cave was probably not inhabited.

CHAPTER SIX

Chronology of the Cave Fill

INTRODUCTION

A major goal of this sedimentological and stratigraphic study is the understanding of the physical history of the site over the period of time when it was occupied by prehistoric people. The sequential development of the sedimentary fill in Franchthi Cave can be reconstructed on the basis of the information in the preceding chapters, namely, (a) description of the individual strata, or lithostratigraphic units, (b) interpretation of the laboratory analyses in terms of origins of the primary sediment and recognition of unconformities (hiatuses) identified by weathering phenomena, and, in this chapter, (c) the radiometric control provided by radiocarbon dating. First, the radiocarbon data will be reviewed; next, the impact of radiocarbon dating on trench-to-trench correlation will be examined. Then those data will be plotted as a function of stratigraphic depth in order to evaluate the rate of sediment accumulation. Combining the above results with the evidence of weathering, one can reach conclusions about the presence and duration of sedimentary hiatuses—unconformities—in the stratigraphic sequence. The tephra of Stratum Q is of major importance as a regional chronostratigraphic marker as well as a key marker bed in the cave, and its age will be reviewed below. The chapter will conclude with a discussion of the changes in the configuration of the bedrock cavity followed by an interpretation of the evolution of the cave infilling.

RADIOCARBON DATING

The radiocarbon determinations are tabulated in Table 6.1, modified from Plate 71 of Jacobsen and Farrand (1987), which listed 59 radiocarbon dates from the cave, plus one thermoluminescence date on potsherds that agrees with the enclosing radiocarbon dates. Five other radiocarbon dates were determined on charcoal recovered from a submerged site offshore from Paralia, but no material for numerical dating was recovered from the Paralia excavations. The offshore dates range from 6200 to 7600 B.P., supporting the idea that the offshore site was originally part of the Paralia settlement (Gifford 1990).

   Almost all the radiocarbon analyses were made on charcoal; the few exceptions are listed as "carbonized matter" or, in one case, "charred bone." The ages discussed here are reported in terms of "B.P.," "Cal B.P.," and "Cal B.C." "B.P." means "years before A.D. 1950 based on the 5568-year half-life of ^{14}C," which is the approved convention of all radiocarbon laboratories. "Cal B.P." indicates the calibrated age of the "B.P." date—always in years before A.D. 1950—derived from dendrochronology and thus represents the best estimate of true calendrical age.

Table 6.1. Radiocarbon Dates from Franchthi Cave.

EXCAVATION UNIT	STRATUM	UNCAL B.P.	STAND DEV	CAL B.P.	CAL B.C.
A:56	Y2	7190	±110	7950	6000
A:63	X2	7790	±140	8450-8540	6500-6590
FA(QSE):10	Z	105	±40	modern	modern
FA:39	Z	5160	±80	5940	3970
FAN:89	Y2	6110	±90	6955-6990	5005-5040
FAN:97	Y2	6160	±70	7015-7145	5065-5195
FAN:114	Y2	6690	±80	7450-7530	5525-5580
FAN:120	Y2	6855	±190	7635	5625
FAN:129	Y2	6790	±90	7575	5640
FAN:129	Y2	6730	±70	7540	5590
FAN:137	Y2	6670	±70	7480-7530	5530-5580
FAN:177	W3	8710	±100	9650-9780	7700-7830
FAN:177	W3	8730	±90	9650-9810	7700-7860
FAN:177	W3	8530	±90	9490	7540
FAN:197	W2	9150	±100	10045	8095
FAN:218	W2	9250	±120	10,210-10,285	8260-8335
FAS:72	Y3	5260	±60	5990-6030	4040-4080
FAS:83	Y2?	6170	±60	7025	5075
FAS:102	Y2	8410	±90	9390-9430	7440-7480
FAS:129	Y1	6940	±90	7710	5760
FAS:143	X2	7930	±100	8665-8945	6715-6995
FAS:146	X2?	7980	±110	8775-8950	6715-6830
FAS:195	W1	9430	±160	10,375-10,420	8425-8470
FAS:195	W1	9060	±110	9990-10,020	8040-8070
FAS:197	W1	9210	±110	10,090-10,270	8140-8320
FAS:197	W1	9280	±110	10,210-10,290	8260-8340
FAS:204	U	10,260	±110	11,880	9330
FAS:207	U	10,840	±510	12,600	10,650
FAS:207	U?	14,680	±990	17,360	15,410
FAS:209(=207)	U	10,790	±160	12,540	10,590
FF1:W=7.45 m/sl	Y2	6750	±80	7550	5600
FF1:W=6.7 m/sl	Y2	6830	±60	7620	5670
FF1:42B1	Y2	7700	±80	8380-8410	6430-6460
FF1:44B5	X2	7900	±90	8610-8700	6660-6750
FF1:43A1	X2	8020	±80	8960-8985	7010-7035
G:31	X2?	9100	±140	10,000-10,035	8050-8185

Table 6.1. (Cont.)

EXCAVATION UNIT	STRATUM	UNCAL B.P.	STAND DEV	CAL B.P.	CAL B.C.
G1:11	Y2	6650	±80	7470-7530	5520-5580
G1:22	X2	8190	±80	9045-9190	7095-7240
G1:39	W2	9030	±110	9990-10,010	8040-8060
G1:46	W2	9840	±100	9955	8005
G1:46	W2	8720	±110	9660-9825	7710-7875
G1:60	W1	9260	±140	10,210-10,290	8260-8340
H:3Y	Y1	7280	±90	8010-8100	6060-6150
HA:59A1	T2	11,090	±260	12,910	10,960
HB:61B1	W2	9300	±130	10,215-10,295	8265-8345
HB:71B2-3	S2	11,930	±170	13,950	12,000
H1:22	Z	40	±40	modern	modern
H1A:101	X2	8940	±120	9955	8005
H1A:117P	W3	9480	±130	10,485-10,530	8535-8580
H1A:117R	W3	8740	±110	9660-9820	7710-7870
H1A:173	U	10,460	±210	12,130	10,180
H1A:175	T3	10,880	±160	12,650	10,700
H1A:181	T2	11,240	±140	13,100	11,150
H1A:199	S2	12,540	±180	14,710	12,760
H1A:219	R	22,330	±1270	26,850	24,900
H1B:126	W3	9290	±100	10,215-10,290	8265-8340
H1B:139	W2	9300	±100	10,215-10,295	8265-8345
H1B:139	W2	9270	±110	10,210-10,290	8260-8340
H1B:191-192	R	21,480	±350	25,800	23,850

Note: This list is modified from Jacobsen and Farrand (1987, Plate 71), but the dates are presented in the same order with a one-sigma standard deviation. The calibrated dates have been revised and extended on the basis of calibration curves in Stuiver and Pearson (1993), Pearson, Becker, and Qua (1993), Kromer and Becker (1993), and Bard et al. (1993). The range of calibrated dates indicates the minimum and maximum intercepts on the calibration curve for a given uncalibrated date. In some cases there may also be intermediate intercepts as well. The calibrated dates indicated here do not take into account the standard deviations of the original date or of the calibration curve. Stratum assignments follow Table 4.1 herein; a question mark indicates a problematic assignment because of crosscutting. See Jacobsen and Farrand (1987, Pl. 71) for information on the laboratory number, material dated, pretreatment, and original place of publication of the dates. In all cases, "B.P." stands for "years before A.D. 1950."

"Cal B.C." is simply the Cal B.P. age minus 1950 years. The sources of calibration curves are given in Table 6.1.

The spread of Cal B.P. dates for a given B.P. date results from the multiple intercepts of the B.P. age with the wiggles of the dendrochronological curve. I have listed only the oldest and youngest intercepts for convenience, but for some B.P. dates there may be as many as four or five intercepts, all of which are equally probable in the absence of other, independent information. Moreover, I have not included the total possible range of Cal B.P. ages for a given sample, which would include the standard deviation of the B.P. date and the standard deviation of the calibration curve. Nevertheless, that statistical uncertainty should be kept in mind.

It should be pointed out that the charcoal samples collected for radiocarbon dating were normally handpicked by the excavators in the trenches, although a few samples came from the water-sieving operation. Charcoal collected from the trenches was commonly in the form of small pieces dispersed across the surface of the square being dug, and it cannot be ascertained that all the pieces came from a single piece of wood. These small pieces were bagged together and labeled with their excavation unit, but no coordinates were measured. Thus, a radiocarbon date cannot be pinpointed in the stratigraphy with any greater precision than that of its excavation unit; see Chapter 2 for concerns about excavation units. In a few cases, this is unfortunate because crosscutting has been recognized after the fact, and in these cases it is uncertain which natural layer the sample came from, or if the sample may be a mixture from two or more layers.

In five cases, multiple samples—not repeated analyses of the same sample—were dated from a single excavation unit, with generally good agreement. In one instance, however, a trench sample and a sample from the water sieve were dated for the same unit (H1A:117) and yielded dates 700 years apart. Where there was disagreement, the most appropriate age could usually be selected by reference to dates on surrounding units.

Two radiocarbon analyses were useful in demonstrating that the uppermost sediments (Stratum Z) have been heavily disturbed in very recent times. A hearth in Unit FA:10 in the SE quadrant of FA balk at a depth of 0.70 m below datum produced a date of 105±40 B.P., or A.D. 1845. Another hearth in Unit H1:22 at a depth of 2.64 m below datum was dated at 40±40 B.P., or essentially contemporary, but indicating deep modern disturbance throughout Stratum Z.

INTERTRENCH CORRELATION

As discussed in Chapter 4, the correlation of lithostratigraphic units from trench to trench was based on, as primary criteria, macroscopic features that could be evaluated in the field and on the section drawings, such as key marker beds and the relative sequence of strata. In Chapter 5, the point was made that trench-to-trench correlation was not based on the results of laboratory analyses in order to avoid circular reasoning in the interpretation of those results.

Radiocarbon dating is an independent means of verifying the proposed correlations. Figure 6.1 (modified somewhat from Figure 2 in Farrand 1993) shows the correlations between Trenches FA and HH1 and includes selected radiocarbon ages (in Cal B.P.), especially those closely associated with the boundaries between lithostratigraphic units. The section drawings in Figure 6.1 show specifically the south faces of subtrenches FAS and H1B at a reduced scale; see Figures 4.1 and 4.2 for a better view. The wavy lines in Figure 6.1 indicate the positions of unconformities as interpreted in the field and supported by the laboratory analyses discussed in Chapter 5. The unconformities are of different durations, as we shall see below.

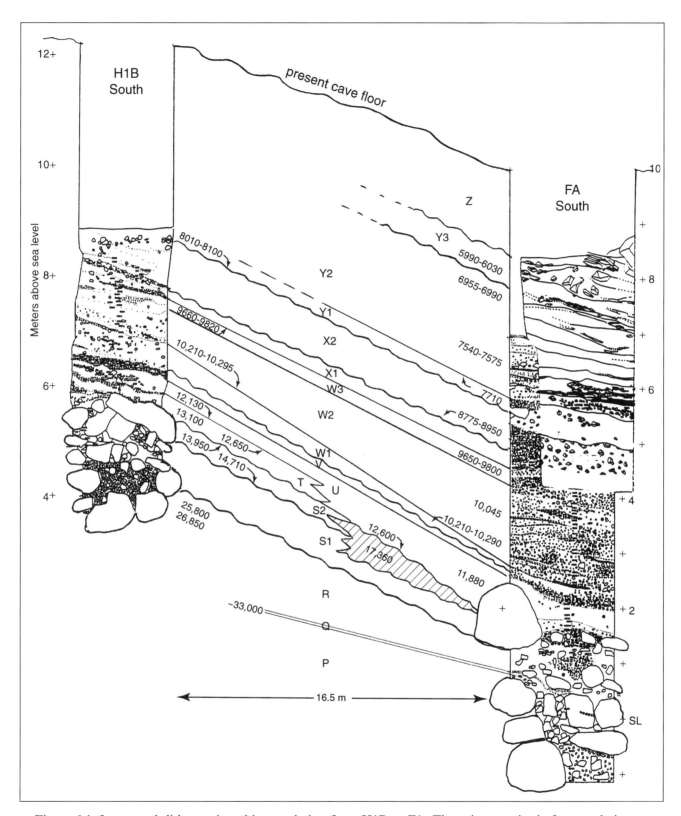

Figure 6.1. Intertrench lithostratigraphic correlation from H1B to FA. The primary criteria for correlation are lithological similarity, sequence, and evidence for unconformities. Secondarily, radiocarbon dates (in Cal B.P.) are used to confirm the correlations. The wavy lines are unconformities (hiatuses). The section drawings are adapted from Plates 9, 15, and 18 of Jacobsen and Farrand (1987).

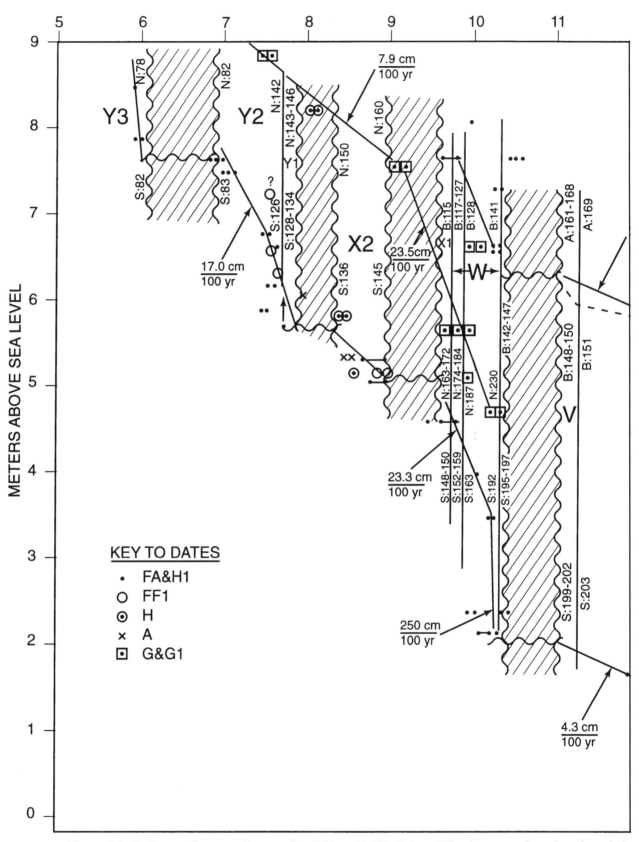

Figure 6.2. Sedimentation rate diagram for 6000 to 18,000 Cal B.P. The dates are plotted against their altitude above sea level. The upper curve is for Trench HH1; the middle curve is Trench GG1, and the lower curve is Trench A-FA-FF1. Separate symbols are used for dates from Trenches A, FF1, GG1,

CAL B.P. x 1000

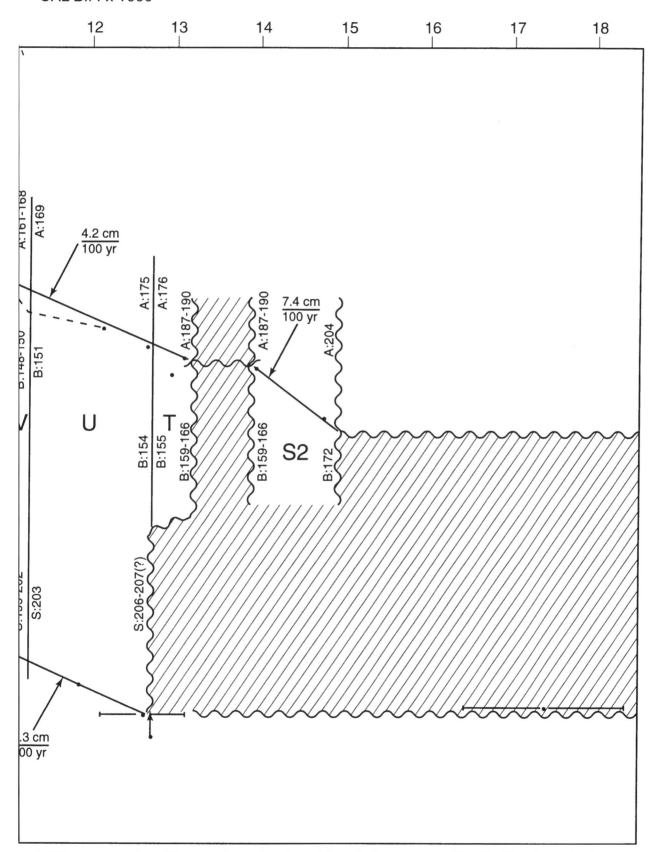

and H because stratigraphic control was less favorable and there was considerable disturbance in those trenches. Bounding excavation units are listed for each lithostratigraphic unit (see also Table 4.1), and hiatuses are shown with diagonally ruled lines. Sedimentation rates, in cm per 100 years, are indicated for representative segments of the curves.

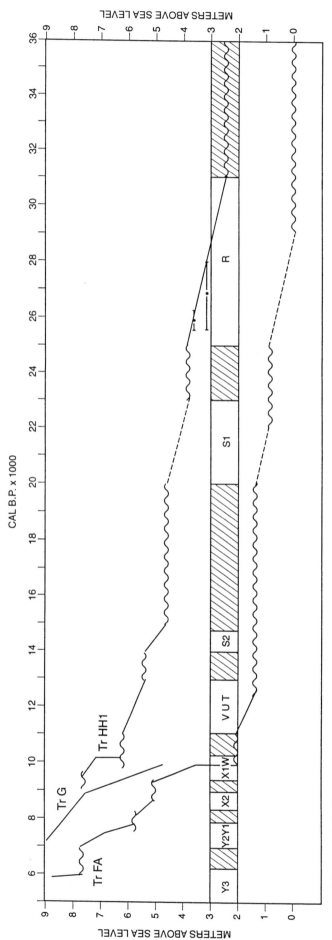

Figure 6.3. Sedimentation rate curves for 6000 to 36,000 Cal B.P. The left half of the diagram is the same as in Figure 6.2. Chronological control for the right half of the diagram is sparse; two dates for Stratum R are shown, but they are statistically identical. There are no dates for Stratum S1. Thus, the dashed portions of the curves are speculative.

The ages shown in Figure 6.1 confirm the proposed correlations in all cases where radiocarbon dates are available. Especially critical for this analysis are the dates associated with Strata U through S1. Field observations led to the conclusion that the shell middens in Stratum T of Trench HH1 are not represented in Trench FA, and it was considered likely that Stratum S2, too, is missing from FA. The radiocarbon dates clearly show that, in Trench FA, Stratum S1 must be older than 17,360 Cal B.P., and the base of overlying Stratum U is about 12,600 Cal B.P., a gap of nearly 5000 years. On the other hand, the dates from Strata T and S2 in Trench HH1 range from 12,650 Cal B.P. to 14,710 Cal B.P. and, thus, fit neatly into the missing time interval of Trench FA.

Some explanation is necessary concerning the dates just mentioned above. On one hand, in the complete listing of radiocarbon dates (Jacobsen and Farrand 1987:Plate 71), a date of 12,540 Cal B.P. is erroneously listed as coming from Unit FAS:209, nominally in Stratum R. That sample, however, was collected from a very clear pit below Unit FAS:208 and intrusive into FAS:209. According to its date, that sample appears to relate to Unit FAS:207 of Stratum U. On the other hand, two dates are reported for Unit FAS:207—12,600 and 17,360 Cal. B.P. in the preceding paragraph—that appear to be clearly at odds. My interpretation of these ages places the 12,600 Cal B.P. date into the basal part of FAS:207 and the older date into the large time gap between FAS:207 and FAS:208, in other words between Strata U and S1. The 17,360 Cal B.P. date has a large standard error (one-sigma) of ±990 years because it was a very small sample, but even so it is statistically distinctly older than the other date for FAS:207 and distinctly younger than the dates for Stratum S2. I believe that this is a valid date, indicating human activity somewhere in the cave at that time. It is conceivable that some activities in other areas of the cave that we did not excavate could have introduced the dated material in the Trench FA area, which appears otherwise to have seen very little sedimentation at that time. In any case, it provides a limiting date for the hiatus separating Stratum S2 from Stratum U in Trench FA.

SEDIMENTATION RATE

The next step in the chronological analysis is to plot the radiocarbon dates against altitude above sea level in order to evaluate the rate of sediment accumulation. A preliminary analysis of this sort was presented a few years ago (Farrand 1993:Figure 4), and the more thorough analysis that follows has not changed the earlier conclusions to any significant extent, but the dates have been converted to Cal B.P. and are therefore older.

Construction of the Curves

In Figures 6.2 and 6.3, the dates are plotted separately for Trenches FA, GG1, and HH1 because of the different vertical dimensions of each of those trenches. Figure 6.2 covers only the time span from ca. 19,000 to 6000 Cal B.P., while Figure 6.3 covers the entire sequence postdating the Tephra Q, but in less detail. The primary control for these curves comes from dates in Trenches FAS, FAS, H1A, and H1B, which are the best documented by excavators' notes; these are shown by single black dots. Supplemental dates from Trenches A, FF1, H and H Ped have also been plotted with different symbols. They agree with the main series of dates in eight of eleven cases. In two instances dates from intrusive features (FAS:"209," and FAS:129) are shown with arrows indicating that they relate to a stratigraphically higher position than that indicated simply by their height. In cases of a date with an unusually large standard deviation (>±160 years), the standard deviation is shown with a solid

horizontal line. Inconsistent dates usually mean that either the date is in error or its stratigraphic position is in doubt.

The dates were plotted according to their normalized heights (see "Presentation of
Data" in Chapter 5), and then they were connected (by eye) by straight lines spanning as
many dates as could reasonably be accommodated with minimum scatter. All straight-line
segments are based on at least two dates. In one case as many as six dates could be lined up,
e.g., in Strata Y1 and Y2 in Trench FA, although three other dates in that span were
rejected in order to draw as conservative an interpretation as possible. It is an area in
Trenches FA and FF1 with considerable disturbance (Vitelli 1993), which could explain
mixed provenience of the charcoal.

The result is a series of straight-line segments for each trench, each with a slope different
from that above or below, or with a similar slope but offset from the adjacent segment(s).
The offsets indicate an interval of time during which there was no net accumulation, in other
words, a hiatus in sedimentation, or an unconformity in geological terms. These hiatuses are
shown by wavy horizontal lines separating the straight-line segments. In cases where there is
a change in slope without a hiatus between two segments, a change in sedimentation rate is
indicated, but the point in time for that change is not necessarily exactly at the date shown,
e.g., the sharp change in slope near the base of Stratum W2 in both FA and HH1. Additional
samples are required in order to confirm the exact position of that inflection point.

Interpretation of the Diagrams

From Figures 6.2 and 6.3, it can be seen that sedimentation rates from the final Upper Palaeolithic (Stratum T and upwards) varied from <10 cm to as much as 250 cm per 100 years. For
much of that time, the rate was between 4 and 17 cm per 100 years. It was only during the
lower and middle Mesolithic occupations (Strata W1 and W2) that the rate increased dramat-
ically, presumably as the result of intense human activity, including the accumulation of snail-
shell middens. In contrast, it is interesting that during the episode characterized by the Upper
Palaeolithic shell middens in Stratum T, the sedimentation rate was not particularly high.
There are only two dates for control in that interval, but the surrounding dates in Strata S2
and U do not allow much deviation from the rate shown on Figure 6.2.

The interpretation for the Final Neolithic of Stratum Y3 is not particularly firm. Both
the sedimentation rate and the length of the preceding unconformity depend entirely on the
date for Unit FAS:72. The rate seems too high compared to that of the preceding Neolithic
strata, and the unconformity appears unreasonably long. Analytical data to support an
unconformity are present, but they are not strong.

Radiocarbon data for Trench GG1 are sparse, but they indicate general trends parallel-
ing those of the two major trenches. In particular, the high sedimentation rates in Strata W1
and W2 (Lower and Middle Mesolithic) are well documented in Trench GG1. Only one date
from GG1 was rejected, namely a date of 10,000 to 10,035 Cal B.P. from unit G:31 in Stratum
X2, the age range of which is firmly placed between 8950 and 8300 Cal B.P. in Trench FA.
Vitelli (1993) has indicated deep disturbance in Trench GG1, which presumably explains this
unacceptable date.

Sedimentation rates for the lower strata (below S2) are largely indeterminate. The only
radiocarbon dates available are those from Stratum R, which have very large standard deviations and statistically could be of the same age, ca. 25,800 to 26,850 Cal B.P. The age of the
Stratum Q tephra (ca. 33,000 to 40,000 calendar years, see below) can also be taken into
account to provide a general idea of the rate of accumulation. Within these limits, an average
rate for the interval from the tephra Q to the base of Stratum R is about 0.5 cm per 100 years;

base of Stratum S1, about 2 cm per 100 years; and the overall average from the tephra to the base of S1 is about 1.25 cm per 100 years. I must stress that such an average figure hides much detail. There are certainly hiatuses in that part of the sequence, such that rates for certain intervals may have been faster than 0.5 to 2 cm per 100 years. On the other hand, much of the mass of the sediment from Stratum S2 to the base of the excavation is in the form of large limestone blocks tens of centimeters thick that must have fallen instantaneously. This implies that the fine sedimentary matrix accumulated—on average—even more slowly than the average figures cited above. The general conclusion to draw from this discussion is that the sedimentation rate prior to ca. 12,000 B.P. was considerably slower than after that date.

Hiatuses

Some of the hiatuses suggested by the plot of radiocarbon dates have been confirmed by the laboratory analyses discussed in Chapter 5; see Figure 5.2. This is very clear in the case of the Y3/Y2, X2/X1, W1/V, and T1/S2 unconformities, as indicated by reduced $CaCO_3$, high index of roundness, increased porosity, white rocks, and chalky snail shells. Surprisingly, the analyses do not reveal a weathered horizon at the S1/R unconformity, but they point to a sharp change in sediment type at that level from Stratum S1 with chalky shells and some reduction of $CaCO_3$ along with increased porosity to Stratum R without those characteristics but with abundant stalactite fragments and shiny snails. This unconformity may have experienced some erosion that removed evidence of weathering at the top of Stratum R, although the field evidence is equivocal on this point.

Table 6.2. Durations of Strata and Hiatuses in Trenches FA and HH1 in Calibrated Years

STRATUM	DURATION	HIATUS IN HH1	HIATUS IN FA
Y3	>200		
Y2/Y3		?	800?
Y2	1200		
Y1	200		
Y1/X2		?	450
X2	650		
X2/X1		?	600
X1	150		
W1,2,3	650		
W/V		650	650
V	250		
U	1400		
T	550		
T/S2		700	n.a.
S2	1000		
S2/S1		ca. 5000 ?	n.a.
S1	1500–2000 ?		
S1/R		2000–2500 ?	2400 ?
U/S1 (in FA)		n.a.	7300
TOTAL Y2 - S1/R	7550–8050	>8350–8850	11,400

The importance of hiatuses can be seen clearly in Figure 6.4, which is modified from Farrand (1993:Figure 5) and now cast in calibrated B.P. ages. The durations of both the strata and the hiatuses are listed in Table 6.2. The total span of time represented by the strata in Table 6.1 from the top of Y2 to the top of R (or the base of the S1/R unconformity) is ca. 19,000 years. Thus, only 8050 years, or 42%, of that span is represented by sediments, and even less than that in FA, where Strata T and S2 are missing.

Vitelli (1999) has suggested that other hiatuses may be present, namely, during her FCP 3/4 interphase (just below unit FAN:110) and during FCP 2/3 interphase (units FAN:119–121 and FAS:116–117). The latter may have lasted "a generation or less," according to Vitelli. These hiatuses are interpreted on the basis of very sparse occupation debris in the sediments that accumulated during those intervals, but they are not hiatuses in sedimentation. Some sediment was still accumulating, and nothing in the sediment analysis indicates a significant pause at those times. Nevertheless, a strong peak in coarse rock fragments (32–128 mm in units FAN:120–122; see Figure 5.2 at 6.5 m above sea level) coincides with the FCP 2/3 habitation hiatus, indicating accelerated rock fall at that time that may have had an influence on the inhabitants.

Volcanic Tephra

The tephra, or volcanic ash, deposit of Stratum Q records an interesting, but brief moment in the filling of Franchthi Cave. It appears to be a primary air-fall deposit, at least in Trench FA, as judged by its purity, excellent size sorting, sharp lower contact, and thickness. In Trench HH1, the tephra is somewhat dispersed, perhaps owing to reworking on the cave floor after its initial deposition. Volcanic ash in primary position is an excellent marker bed for stratigraphic reconstruction. It represents a very brief episode in real time, and an essentially instantaneous moment in geological time, and the tephra is usually distributed over a wide geographic region, making it very useful in correlating events in disjunct areas. Even within Franchthi Cave, the tephra was essential in establishing the correlation of the lower beds in Trenches FA and HH1, given the absence of radiocarbon or other numerical dates for those levels.

Deposition of the tephra in the cave seems to imply that the cave mouth was wide open at that time and, perhaps, that one of the "windows" in the cave roof was open, maybe the innermost window, in order to create a draft to pull the airborne ash into the cave. Many excavators at Franchthi remarked in their notebooks that, at certain seasons, quite a strong wind blows through the cave. The dimensions of the cave mouth at the time of the ash fall are uncertain, but it may have been quite a bit larger than now, given that the tephra occurs about 10 m below the present cave floor. The present opening is about 8 m high, so that the opening at the time of Stratum Q could have been twice as high as now. This estimate is uncertain, however, because the history of ceiling collapse prior to the breakdown of the brow during Neolithic times, as demonstrated in Trench H2, is unknown.

The source of the tephra is certainly in the Gulf of Naples area of Italy, as documented by mineralogical and chemical studies and by the mapping of the distribution of the same tephra across the Mediterranean sea floor. The Campanian province is ca. 825 km northwest of Franchthi. The details of the mineralogical arguments are presented elsewhere (Vitaliano et al. 1981; Thunell et al. 1979; Paterne et al. 1986; St. Seymour and Christanis 1995). In brief, the Franchthi tephra matches very closely the mineralogical composition of the Y5 ash found widespread on the sea floor from southern Italy nearly to Cyprus, and Franchthi sits right in the middle of the distribution of that ash. Moreover, the modeled thickness of the Y5 ash in

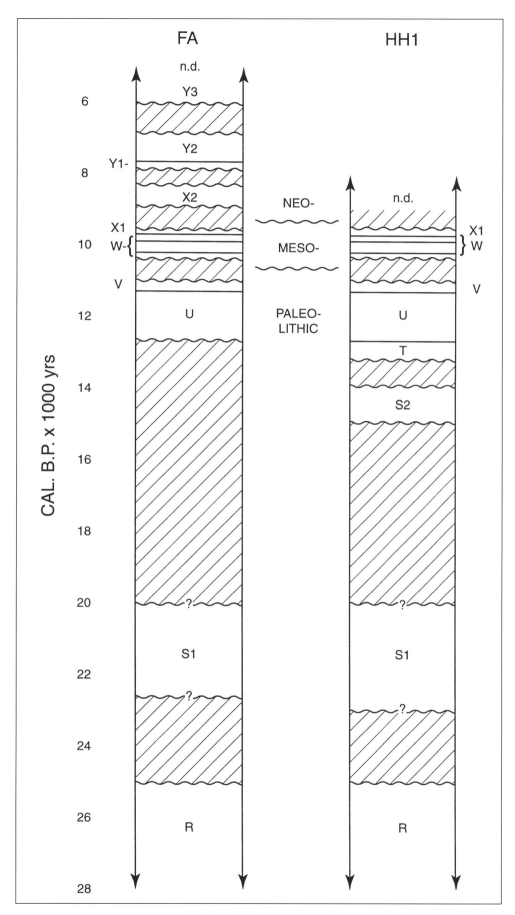

Figure 6.4. Stratigraphic columns showing the duration of lithostratigraphic units and intervening hiatuses. Cf. Table 6.1. Note that the Neolithic/Mesolithic and Mesolithic/Palaeolithic boundaries fall into two hiatuses, each about 300 years long.

the vicinity of Franchthi is ca. 7 cm (Cornell et al. 1983), which matches the thickness of the tephra in Trench FA, ca. 5 to 9 cm.

The Y5 ash is not dated directly, but Paterne et al. (1986) have estimated its age from detailed study of oxygen-isotope zonation in deep-sea cores between Naples and Sicily. They believe the Y5 ash is the same as their C13 ash, which they correlate mineralogically with the Citara eruption on the isle of Ischia dated by K/Ar at 40±2 ka (dated by Gillot in his 1984 thesis [not seen] and cited by Paterne et al. 1986:277). (The symbol "ka" means thousands of years ago and is approximately equivalent to "Cal B.P." for radiocarbon data.) However, Gillot et al. (1982) give a K/Ar date of 33±2 ka (an average of two dates) for the Citara eruption. A further complication is the correlation of the Y5 ash, which more commonly has been matched with the Gray Campanian ash of the Naples region by Thunell et al. (1979), by Cornell et al. (1983), and by us (Vitaliano et al. 1981). The Campanian ash has been radiocarbon-dated on land between 30,000 and 36,000 Cal B.P. (calibration after van Andel 1998a), with the possibility that the younger of those dates might be contaminated, and by K/Ar at 33±1.5 ka (see Paterne et al. 1986:275).

(It is important to recall that radiocarbon dates in this time range are younger than true calendar ages, for example, conventional radiocarbon ages are about 3 thousand years too young at 30 ka [Bard et al. 1990]. K/Ar dates, on the other hand, are unaffected by the cosmic events that produce deviations of radiocarbon ages and thus are better approximations of calendar ages.)

More recently, St. Seymour and Christanis (1995) have reported on a tephra deposit in a small lake in the Kalodiki basin in Epirus, about 300 km northwest of Franchthi. That locality lies nearly on a straight line connecting Franchthi and the Campanian area. Peat just 10 cm above the tephra is dated at 31,800±1200 B.P. (ca. 34,500 Cal B.P.: van Andel 1998a) Mineralogical analyses of the Kalodiki tephra show "strong similarities to tephra from the Campanian province" (St. Seymour and Christanis 1995:46, Tables 1 and 2), and comparison of their analyses with ours (Vitaliano et al. 1981) shows an excellent match. This occurrence, therefore, reinforces the conclusion that the Stratum Q tephra is part of the Y5 ash fall, stemming from the Campanian area, and its age is somewhere in the range of 33,000 to 40,000 B.P. in K/Ar years. Such an age is reasonable in terms of sedimentation rates in Franchthi, as discussed above, where I assumed an age of 35,000 to 40,000 Cal B.P. for the tephra. It is also reasonable in terms of the artifacts found below the tephra, which hint at a Middle Palaeolithic presence (Perlès 1987).

CHAPTER SEVEN

Evolution of the Cave Filling

INTRODUCTION

To reconstruct the evolution of the filling of Franchthi Cave, one needs to keep in mind several phenomena that were playing roles simultaneously, namely the position of the cave in the landscape, the changing size or shape of the cave, and the mechanisms that were bringing sediment to the cave floor. The last of these three was modulated by changing climate and variable human utilization of the cave. We shall look at these phenomena successively, but keeping in mind that they were occurring together.

THE LANDSCAPE

Although Franchthi Cave now sits just above the Mediterranean shoreline, it was never that close to the sea during the prehistoric occupations revealed in our excavations. A coastal plain some 6 to 7 km wide separated the cave from the sea at the time of the last glacial maximum, ca. 18,000 to 20,000 B.P. (van Andel and Sutton 1987:Figure 13; updated in Jameson et al. 1994: 194–203). The coastal plain shrank as sea level rose during the disappearance of the ice sheets in the northern latitudes, such that sea level was at -53 m at the beginning of the Holocene Epoch, ca. 10,000 B.P., when the cave was still ca. 3 km from the shore (van Andel and Sutton 1987:Figure 15). At the beginning of the local Neolithic occupation, ca. 8000 B.P., sea level was at least 20 m below the present, and the bay that now separates the cave from the village of Kiladha was not yet flooded. Even at the end of the Neolithic occupation of Franchthi 5000 years ago, the sea level was at -7 m, and Kiladha Bay was much smaller and shallower than it is today (Jameson et al. 1994:Figures 3.26, 3.32–3.34).

The coastal plain that separated Franchthi from the sea was traversed by streams, as postulated by van Andel (van Andel and Sutton 1987:Figures 13–17), that must have been carrying sediment derived from the hinterland to the east and southeast of the cave. That alluvial sediment comprised eroded soil and rock detritus derived from a variety of bedrock types, as described in Chapter 3 and mapped by Vitaliano (Jacobsen and Farrand 1987:Plate 1; see also van Andel and Sutton 1987:13). Limestone debris derived from the Franchthi limestone was included, of course, but also a mixture of rock types and minerals stemming from the ophiolite complex that underlies the Fourni valley as well as the coastal plain itself. Recall that the freshwater springs just below the Paralia appear to emerge at the contact of the Franchthi limestone with the underlying ophiolite rocks, which presumably continue below the waters of Kiladha Bay off Paralia. Moreover, many other submerged springs occur along the present shoreline of the large bay just north of Franchthi (van Andel and Sutton 1987:Figure 5), probably along the same bedrock contact, and they would have been exposed and available to the local people whenever sea level was lower than now.

Several generations of stream alluvium can be observed in the Franchthi area and throughout the southern Argolid, and similar deposits must occur on the drowned coastal plain. This alluvium was undoubtedly a major source of the exotic rocks and minerals found in the cave sediments that were brought in largely by human activities, intentional or not, as discussed in Chapter 5 under "Sediment Origins."

In their study of the southern Argolid, Pope and van Andel (1984) concluded that episodes of alluviation occurred both before (Loutro alluvium >32,000 B.P.) and after (Pikrodafni alluvium <4500 B.P.), but not during, the documented occupation of Franchthi Cave. They hypothesize that, during the time represented by those occupations, "all streams seem to have been deeply incised, a condition compatible with moderate rainfall and well-vegetated slopes" (Pope and van Andel 1984:296). One may speculate, however, that the soil cover on the mountain slopes was thin at that time, given that thick alluvium and debris flows of the upper two members of the Loutro alluvium, both of which contain Middle Palaeolithic artifacts, must have stripped the bulk of the former soils from the hill slopes.

Excavators mentioned "polished sea pebbles" or "small water-washed pebbles," as well as "sea sand," in several excavation units, mostly in Mesolithic Strata X2 (units FAS:136, FAS:143, FAS:145, and H1A:85), W3 (units FAS:151 and 157), and W2 (unit FAN:198), but these pebbles never occurred in abundance. A few of them appeared in two or three of my samples, in particular FR 1–21 and 1–23, from Strata X1 (units FAN:165 and 167) and W3 (units FAN:172–174). These pebbles could just as easily be stream pebbles, collected by the inhabitants of the cave for whatever reason, although the abundance of marine shells and large fish bones in some of the same units does indicate that the people were frequenting the sea shore, which was only 2 to 3 km distant from the cave at those times.

In summary, the landscape around Franchthi Cave in prehistoric times was rather different from that which we see today. An extensive coastal plain separated the cave from the sea, and it was traversed by streams bringing a variety of mineral and rock components into the area. The upland slopes were well vegetated, and the streams were incised into the plain, implying that the coastal plain was probably well vegetated, too. Freshwater would have been available from a number of springs now submerged. Marine resources would have been available within a one- to one-and-a-half-hour walk (or less) from the cave.

CHANGING CONFIGURATION OF THE CAVE

There is much evidence that the cave has changed in size and shape during the time represented by our excavations, but the overall proportions probably have not changed in a major way. The length and width of the bedrock chamber ca. 35,000 Cal B.P. were undoubtedly nearly the same as today. Possibly the brow of the cave extended a bit farther toward the Paralia, judging from some large collapsed blocks encountered in Trenches B and E, now 5 to 12 m outside the present dripline, but the age of that collapse is not known. The pile of breakdown under the current brow and encroaching into Trench H2 indicates that the front of the cave extended at least a few meters farther out in Neolithic times.

The deposits under the window at the inner end of the cave are not dated, but they appear older, in my judgment, than do those at the brow or those under the big breakdown window in the middle of the cave. This cannot be substantiated, however, at the present state of our knowledge.

Information is limited concerning the width of the cave through time. It is known that the area of the pond in the rear of the cave was available for human activities throughout the time represented by our excavations, but it is not known if the configuration of the cavity containing the present-day pond changed with time. Only in the case of Trench GG1 can the cave wall be observed below the present floor. In that case, the bedrock wall expands outward with depth, so that at a depth of 5.92 m below datum (5.40 m above sea level), the western cave wall is 2.5 m farther west than at the present surface (Figure 2.5). It is not known if this trend continues with greater depth, or if the same expansion applies to the eastern cave wall. Recall, however, that the electric resistivity survey (Chapter 3) suggests that the cave fill may extend at least 5 m below present sea level, that is, 10 m below the lowest observed point on the cave wall. On the other hand, there is a natural limit to the cave depth, namely the level of the contact between the cave limestone and the underlying ophiolitic rocks. That contact is about 1 m below sea level at the shore in front of Franchthi, but its configuration is not known. It must dip to a lower level under the cave proper, if the resistivity results are correct.

The topography of the cave floor appears to have varied from its present configuration in the past. Today the floor is highest in the area of Trench HH1, lowest around Trench FA, and at an intermediate level at Trench GG1. Table 7.1 shows how those relations have changed through time.

Table 7.1. Topography of cave floor through time (in meters above sea level)

TRENCH	GG1–SE Corner	FA–NW Corner	HH1–NW Corner
PRESENT FLOOR	11.3	10.0	12.3
NEO/MESO LIMIT (X2/X1)	6.6	5.3	8.3
MESO/PALAEO LIMIT (W1/V)	3.3	2.2	6.0
TOP OF S1	?	1.75	4.75

Thus it can be seen that the slope from GG1 to FA has been maintained through time at ca. 1-m drop across a distance of ca. 12 m as one goes into the cave. In contrast, the differences between GG1 and HH1 and between FA and HH1 have decreased through time. From the time of Stratum S1 through the Mesolithic occupations, that difference was 3 m or more for FA vs HH1, compared to 2 m at present. The difference between GG1 and HH1 was in excess of 2.5 m, whereas it is only 1 m today. Therefore, the area of Trench HH1 was always the high point on the floor in the excavated part of the cave, although the cave floor on the east side of HH1 may have been higher in the past, judging from the strong dip of the strata from east to west as seen in the excavations of HH1 (e.g., the south section of Trench H1B in Figure 4.2). Conversely, the area of Trench FA was noticeably lower, and perhaps damper, probably leading to differential use of space in the cave. Recall that the Palaeolithic snail middens (Stratum T) do not occur in the FA area.

The collapse of the ceiling of the cave is quite obvious from the huge breakdown pile in the middle of the cave and the large window to the sky that resulted. That collapse appears to post-date the Neolithic occupations in the cave, as discussed in Chapter 3, and it seems likely that the cave floor was relatively flat during those occupations. In addition to the excavations in the front part of the cave and the small test excavation near the rear, where the Neolithic levels occur at ca. 9 to 12 m above sea level, there is a "chasm" within the breakdown pile, the bottom of which is now at ca. 15 m above sea level (Figure 3.2, Plate 4a). It is filled with some 2.5 to 3 m of loose sediment, lying on immovable limestone blocks, which are, thus, at about the same level as the top of the Neolithic levels elsewhere.

Prior to this catastrophic collapse, some detachment of large blocks up to 1 to 2 m across from the cave ceiling was occurring. This is especially clear in the lower strata from S2 downward (Figures 4.1 and 4.2), but occurred occasionally during the accumulation of the upper strata. Such roof fall means that the ceiling of the cave was being raised progressively with each block that fell, but this was most likely a spatially random process, with a block falling first here and then there through time. Given the unstable structure of the bedrock where the bedding planes are nearly vertical and the rock has been shattered by past tectonic movements, groundwater solution was the likely agency that detached the large blocks. The abundance of these large blocks in the lower strata implies that the ceiling height was increasing somewhat rapidly during Palaeolithic times, but was, on the whole, relatively stable during the Mesolithic and Neolithic occupations. The height from the sedimentary floor to the ceiling probably decreased over time, however, given that the finer components of the fill—loams, anthropogenic imports, etc.—came from outside the cave, while only the limestone fragments and large blocks came from the ceiling. In other words, the total volume of the sedimentary fill is greater than the volume of the limestone roof-fall components of that fill, and thus the top of the fill was rising faster than the ceiling was.

As described above (Chapter 5), the layers of *éboulis secs* in Strata V and X1 seem to call for a different mode of detachment, perhaps earthquakes, that brought down abundant, small fragments of the ceiling rock. Volumetrically these thin beds of *éboulis secs* would account for only a slight raising of the cave ceiling, relative to the collapse of meter-sized blocks.

In summary, the size and shape of the cave probably have not changed to a great extent over the past 35,000 years, except for the opening of the large windows to the sky. The length of the cave has probably stayed about the same over that period of time, while the width may have decreased somewhat, perhaps 5 to 10 m, through time as the sedimentary fill accumulated against bedrock walls that converge upward, as was documented in the case of Trench GG1. The height of the ceiling above the sedimentary floor decreased through time because more sediment was introduced into the cave than bedrock fell from the ceiling. The topography of the floor of the outer chamber changed throughout this time. The area of Trench HH1 was always highest, but in Palaeolithic and Mesolithic times it was even 1 to 1.5 m higher relative to the areas of Trenches GG1 and FA, perhaps leading to different functional uses of each of these areas than would be predicted on the basis of the present configuration.

The opening of the large window in the middle of the cave obviously brought about a major change in the conditions inside the cave, but this occurred at or after the end of the Neolithic occupation. Prior to that collapse, the interior of the cave would have been very dimly lit by natural light, or even quite dark if the window at the far end of the cave was not yet open. Thus, it is likely that the interior of the cave was dark during most of the time represented by the occupations revealed in our excavations, although the excavated area was well lit owing to the large size of the cave mouth.

Still more speculatively, it is possible that the Palaeolithic inhabitants might have had access to a much longer cave passage. The floor of the pool at the back of the cave is 2 to 13 m below present sea level and connects with farther passages that were not explored (Chapter 3). Therefore, about 35,000 to 40,000 years ago, when Strata P and Q were accumulating in the front part of the cave and sea level was well below its present level (approximately -80 m), the cavern in what is now the pool area would have been dry and accessible.

EVOLUTION OF THE INFILLING

The sediments filling Franchthi Cave change character through time as a function of the interplay of natural processes and the intensity of human occupation. As was demonstrated above, prior to ca. 13,000 or 13,500 Cal B.P., natural processes were dominant, and human influence seems to have been minimal. Accordingly, the rate of sedimentation was much slower before that date than it was afterwards. We will follow the changing character of the sediment from the base of the excavations to the top, primarily as seen in Trenches FA and HH1. Recall, however, that our view of the sediments in the cave is limited in scope in two ways. First, our excavations examined only about 10% of the cave floor in the outer part of the cave. Secondly, the resistivity survey suggests that there may be at least 5 m more of sedimentary fill below the deepest point that we reached. Thus, our sample is restricted, and there is much of the cave that has not been seen.

The Basal Strata P, Q, and R

Strata P, Q, and R constitute the earliest component of the cave fill. Strata P and R are very similar yellowish red sandy clay loams deposited on and between large limestone blocks. They are separated stratigraphically by the tephra of Stratum Q, but their genesis is otherwise the same. There is a conspicuous amount of exotic sand grains in these loams, such that the excavators commonly remarked on their sandy nature. Those observations must be tempered for Stratum R, however, which contains reworked, fine-sandy tephra grains that would enhance the sandiness of the sediment. The large roof-fall blocks are more numerous than in any overlying strata; in fact, they became impenetrable at the base of Trench HH1. However, these roof falls appear to be different from that under the large window, in that the blocks are separated by fine matrix with smaller rock fragments and the presence of human activity. It is likely that they fell one by one, or at most only a few at a time.

The origin of the sediment of Strata P and R must have been mainly the reddish brown residual soils (terra rossa) that formed on the bedrock outside the cave and subsequently were washed or filtered into the cave through fissures in the bedrock. Whether these fissures were located in the part of the cave that can be seen today or elsewhere on the Franchthi upland is not known. As mentioned in Chapter 3, some sinkholes and a deep karstic shaft are known on the upland. The large windows that exist at present could have begun as much smaller openings, but that will never be known. The possibility exists that the yellowish red clays could have entered the cave system through the passage that debouches at the pool in the rear of the cave. That pool is floored with "fine red silt," according to divers, which could be the equivalent of the loamy matrix excavated in the outer part of the cave. The excavators noted that the yellowish red clay loam was quite different from that they had dug above Stratum R, not only in color, but also because it was noticeably wet and sticky. Another difference was the conspicuous presence of stalactite fragments in these sediments, probably indicating considerable roof drip in a moister regime. The exotic sand component of these loams can be attributed, at least in part, to human agency. Chert artifacts or debitage, fragments of large mammal bones, and marine shells occur throughout, especially in Stratum R. Given that sea level was much lower at that time, the inhabitants would have been walking back and forth across the coastal plain, which is underlain by rocks of the ophiolite complex, and dragging some of this rock debris into the cave. There is also the possibility, even the likelihood, that some of the fine-grained sand grains may have been blown in by the wind sweeping across the Gulf of Argos and the wide coastal plain.

It is conceivable, but untested, that a sizeable unconformity exists between the tephra Q and Stratum R. It was concluded above that the tephra is 33,000 to 40,000 years old, and two dates indicate that Stratum R is ca. 25,000 to 26,000 calibrated-years old. One particular difference between Stratum P and Stratum R is the presence in P of conspicuous light yellow particles, up to 3 cm across, that are decayed fragments of limestone. Their absence in R suggests a distinct difference in age between P and R. Moreover, only about half a meter of coarse sediment separates those two dates mentioned here. If sedimentation were continuous in that interval, the rate would have been exceedingly slow relative to the rates for overlying strata—<0.44 cm per 100 years, after the large limestone blocks are accounted for. Therefore, it is likely that a hiatus of some length follows the deposition of tephra Q.

Strata S1 and S2

Strata S1 and S2 are similar sedimentologically, but disjunct in time. S2 is dated between 14,000 and 15,000 Cal B.P. in Trench HH1, but is absent in FA. S1 is older than 17,360 Cal B.P. in Trench FA, and I have estimated that the S1/S2 hiatus in Trench HH1 may be as long as 6500 years (Table 6.2). Both of these strata are characterized by large roof-fall blocks, some in excess of one meter in length and many in the 50 to 60 cm range, sitting in a clay loam matrix. There are also numerous smaller, angular, limestone fragments, such that the excavators called certain units "more rock than soil." Stratum S1 has only limited evidence of human activity, but in S2 the anthropogenic influence increases considerably, as does the sedimentation rate. Lithic artifacts increase and diversify markedly in S2 (Perlès 1987:83), and faunal remains—both micro- and megafauna—become moderately abundant in S2, whereas they are almost totally absent in S1 (Payne 1982 and unpublished).

Upper Palaeolithic Middens—Stratum T

The three members of Stratum T contain the first evidence of overwhelming anthropogenic influence on sediment accumulation. Here are repeated lenses of land-snail-shell middens (*escargotières*), and they appear to be most intense in Members T1 and T3 (Figure 4.2). In Trench H1B the number of snail shells reaches 700 to 1200 in a one centimeter slice across some excavation units (Whitney-Desautels forthcoming). In some middens, there is almost no sedimentary matrix among the snails, most of which are whole shells of *Helix figulina*, but there is abundant charcoal, animal bones, and chert artifacts. Fish bones occur, and marine shells appear for the first time, and they are moderately abundant (Shackleton 1988:Figures 5, 6). A few of the land-snail shells and some of the bones appear burnt, but on the whole most of the shells do not seem to have been heated directly in a fire.

Strata S2 and T do not appear in Trench FA, and Trench GG1 was not dug deeply enough to expose these strata, if indeed they occur in that area. In Trench FA there was a very long interval, some 7300 years from the top of Stratum S1 to the bottom of Stratum U, without significant sedimentation, and, thus, without significant human utilization of that part of the cave. Recall that the topographic reconstruction described above indicates that the area of FA was even lower relative to the area of Trench HH1 at the time of Stratum T than it is today, and thus it might have been considered an undesirable location by the occupants of the cave. An alternative explanation is that Stratum S1, which is undated and shows almost no human influence, may have accumulated over a much longer time than that suggested in Figure 6.3. If so, the long hiatuses between R and T in Trenches HH1 and FA may have been somewhat shorter than the 7000 and 9700 years, respectively, proposed here (Table 6.2).

The Final Upper Palaeolithic Episode—Strata U and V

Strata U and V may be considered together as a distinctive episode, although each has a distinctive sedimentological character. Stratum U is a rather uniform, reddish to yellowish brown clay loam with few large rocks and limited anthropogenic input. However, the suggested rate of sedimentation is essentially identical to that of Stratum T with all its middens. The excavators noted that the clay of Stratum U was unlike any that had been dug in the overlying strata. The episode is brought to an abrupt close with the deposition of the *éboulis secs* of Stratum V, after which a hiatus of some 650 years occurred. The great abundance of landsnail shells in Stratum T decreases precipitously to only 10 to 20 (or less) per one centimeter slice (Whitney-Desautels forthcoming). Marine shells are absent in many units in Trench HH1 but moderately abundant in FAS (Shackleton 1988). Microfaunal remains continue in moderate numbers, but the megafauna (equids, cattle, deer, and pig) nearly disappears (Payne 1982). In the sediment samples (Figure 5.2), cultural debris is practically nonexistent, and snail-shell fragments drop to less than half of their abundance in T2 and T3. On the other hand, lithic artifacts continue to be moderately abundant throughout Strata U and V (Perlès 1987:83, 1990:Chap. 1). Perlès considers the units in Stratum V to be an interphase (her VI/VII), that is, either a transitional or a stratigraphically mixed lithic assemblage. A possible explanation in this case may lie in the permeable nature of the loose *éboulis,* which would allow artifacts, especially microliths, to filter down from the overlying Stratum W.

The W/V unconformity that follows the Strata U and V episode is well documented by sediment analyses (Figure 5.2), as well as by trench observations. White rocks occur throughout; and high values of the roundness index and porosity, and reduced $CaCO_3$, along with chalky snail shells, mark this hiatus as the most prominent in the cave. The beginning of the hiatus is not well dated, but its end is clearly marked at ca. 10,300 Cal B.P. by an abrupt increase in sedimentation rate in all trenches (Figure 6.2) and the sudden onset of intense human presence, marking the beginning of the Mesolithic cultural period.

Intense Mesolithic Occupation—Stratum W

The three members of Stratum W encompass the Mesolithic occupations of Franchthi Cave. For a period of about 600 years intense human activities augmented the sedimentation rate considerably. In member W1 and the lower part of W2 sediments were accumulating as much as 250 cm per 100 years. Thereafter, the rate decreased markedly to ca. 23 cm per 100 years, which was still four- to six-times greater than in Palaeolithic times. The lowest member, W1, appeared to the excavators somewhat different from the upper two members. The sediment was redder and seemed to have less anthropogenic influence, but, like Members W2 and W3, it was characterized by hearths, ash layers, and crushed snail shells, and was rather rocky. The snail middens occurred in piles, "humps," and pits, as reported by the excavators, along with what were interpreted to be fire-cracked rocks. White rocks, which occurred regularly in the upper two members, were not found in W1, and none of the snails in W1 were chalky.

Member W2, by far the thickest of the three members, is a complex of ash lenses, hearths, snail shells, and abundant cultural debris, including bones of large fish and marine shells, all in a brown to yellowish brown loamy matrix. It was observed that some snail-shell accumulations seemed to lie in troughs. There are also lenses of "rocky red" sediment here and there, and generally throughout Stratum W rock fragments in the 128–32-mm size class constitute 50% or more of the total sediment.

Member W3 is similar to W2, but, on the whole, stonier and with less matrix among the stones. The quantity of snail shells and other cultural debris (Figure 5.2) is somewhat less than in W2, although this may reflect in part a dilution effect resulting from a greater input of natural, rocky sediment. W3 is the stratigraphically lowest stratum reached in Trench H2, just inside the present brow of the cave. Here as in the other trenches, W3 is very rich in crushed snail shells, such that the excavators remarked on the abrupt appearance of shells as they encountered the top of Stratum W3.

Most land-snail shells in all members of Stratum W are crushed, in marked contrast to the whole shells characteristic of the Upper Palaeolithic middens, implying different methods of food preparation practiced in those different times, or different treatment of the discarded shells.

Some white rocks occur throughout Member W2, but the *Helix* shells are mostly shiny, $CaCO_3$ content is high, and roundness is low. These conditions may indicate that weathering conditions were prevalent, but the rate of sediment accumulation was too great for a weathering horizon to have formed. Alternatively, weathering of the white rocks and some snail shells may have been enhanced by the abundance of organic residues, which can degrade into organic (humic) acids. Note that organic matter is consistently high throughout Stratum W (Figure 5.2).

The increased human influence has been noted by several investigators. *Helix figulina* shells again number in the hundreds per one-centimeter sediment slice (Whitney-Desautels forthcoming). Marine shells increase in abundance to 300 to 400 individuals per excavation unit, but the genera change from predominantly *Patella* and *Monodonta* in Strata V and U to predominantly *Cyclope* and *Cerithium* in Stratum W (Shackleton 1988:figures 3–6), implying different shoreline environments. There is a remarkable increase in botanical remains from <100 g carbon per 100 years in all levels prior to the Mesolithic to as much as 6500 g carbon per 100 years in Stratum W of Trench FAS and 2500 to 3000 g carbon per 100 years in Trench HH1. This is essentially the ethnobotanical zone III of Hansen (1991), which is characterized by increasing diversity in vegetal remains. The megafauna reveals the disappearance of equids, only a few cattle, but a dominance of red deer (70% of the fauna) along with pigs (S. Payne unpublished). The lithic analysis of Perlès (1990:114) indicates the same high density of retouched tools as in Strata U and V *per lithic phase,* but the phases were much shorter in duration in the Mesolithic, meaning that the density *per unit time* was much greater during Stratum W.

Stratum X1 and the Neolithic—Mesolithic Gap

Stratum X1 closes the Mesolithic sequence abruptly. These *éboulis secs,* thin in Trench HH1 and up to 30 cm thick in FA, constitute another key marker bed in the Franchthi stratigraphy. It is interesting to note that Stratum X1 appears to be missing in Trenches GG1 and H2, that is, just inside the cave mouth. Elsewhere in the cave, however, this was an episode of rapid deterioration of the cave ceiling, during which human occupation of the cave seems to have declined markedly. Snail shells decrease in abundance (Whitney-Desautels forthcoming), botanical remains drop drastically from their high values in Stratum W (Hansen 1991:zone V), and the megafauna (deer and pig) is scarce (Payne in Jacobsen 1973:figure 6). Retouched lithic artifacts decrease to one-tenth of their abundance in Stratum W (Perlès 1990:114, Phase

IX). Only marine shells seem to maintain moderately abundant numbers (Shackleton 1988:figures 3–6). As suggested above (Chapter 5), the process that triggered this episode of rapid *éboulis* accumulation, whether climatic deterioration or earthquakes, may have discouraged the inhabitants from continuing their use of the cave.

Following the deposition of Stratum X1 sediments, another distinct weathering interval occurred. The effects can be seen in Figure 5.2 by the presence of white rocks in Stratum X1 itself, and by an increase in the roundness index and an accompanying decrease in $CaCO_3$ that begin in X1 and continue down into Stratum W. Porosity is not particularly high, but it is distinctly higher than in the overlying Stratum X2. Although clearly marked, this weathering interval is not as strong as the one in Strata V and U at the end of the Palaeolithic. The duration of this unconformity is estimated to be about 400 to 500 years, ending ca. 8000 B.P. (Figure 6.2). Its beginning is not dated directly because there are no dates from Stratum X1, but the extrapolated sedimentation rate from Members W2 and W3 places the end of Stratum X1 ca. 8500 B.P.

The Onset of the Neolithic Occupations

The hiatus at the end of the Mesolithic was closed when the so-called "gray clay" of Stratum X2 began to accumulate. There is a clear increase in clay-sized sediment relative to sand and fine-gravel components, and rock fragments are relatively scarce. Also, the sedimentation rate is moderately low, about 8 cm per 100 years (Figure 6.2), close to that in Stratum U, where the sediment resembles that of X2 in texture, color, and stoniness. Roundness and porosity are low and very low, respectively, and $CaCO_3$ is high. All of these characteristics point to a very reduced human presence in the cave, and, indeed, cultural debris (Figure 5.2) is decidedly low. However, ceramics appear for the first time, although not consistently in all excavation units (Vitelli 1993:Ceramic Phase 0/1). Retouched lithic artifacts are twice as abundant in this Initial Neolithic phase (Perlès 1990:Phase X) as in the underlying Stratum X1, but still much less abundant than in the Mesolithic strata. Land snails have practically disappeared (Whitney-Desautels forthcoming), and marine shells decrease to quite low values (Shackleton 1988:Figures 3–6). Botanical remains continue at low abundances. Deer continue to dominate the megafauna, although their numbers are much reduced relative to the Palaeolithic of Stratum T, and fish are abundant, totaling 20% or more of the animal bones (Payne in Jacobsen 1973:Figure 6). Sheep and/or goats have not yet appeared in abundance in the cave, if at all. Payne (in Jacobsen 1973:Figure 6) reports none from H1A:101–108, and the few sheep and/or goat bones in G1:20–21 are suspected to be intrusive (Jacobsen 1969b:351–352). Only in Trench A are sheep and/or goat bones reported to be abundant—in A:63–65 (Jacobsen 1969b:352), which I correlate with Stratum X2 on the basis of its sediment type (brown clay with few stones) and a radiocarbon date of 8450 to 8540 Cal B.P. On the whole, Stratum X2 marks a quiet period of steady deposition of pale brown clay loams, little rock fall, and limited occupation of the cave.

The transition from Stratum X2 to Y1 is shown in Figure 6.2 as a minor unconformity of ca. 100 years, based on the slopes for the preceding and following segments of the curve. The laboratory analyses do not add any support to the idea of a hiatus, although such a short period of time would not allow much weathering to be manifested. In any case, there is an abrupt change in sediment type and sedimentation rate beginning in Stratum Y1.

Strata Y1, Y2, and Y3—The Full-Blown Neolithic

These three lithostratigraphic units are treated as full strata, rather than as members, on the basis of contrasting sediment types, although that decision is nevertheless arbitrary. Description and analysis of these strata are not as straightforward as for the underlying strata because they are not well depicted on section drawings, and the excavators did not provide descriptions as detailed as they did for the lower stratigraphic units. Furthermore, there is considerable modern disturbance and mixing in the Y strata, as well as in the succeeding Stratum Z. The degree of disturbance, which varies from trench to trench, is discussed in detail by Vitelli (1993:33 and Plan 2) on the basis of ceramic ethnostratigraphy. Such disturbance is not readily recognizable in the sediments alone, short of finding discrete pits and modern detritus.

Stratum Y1 is a relatively thin, "rocky red" layer. It is as much as 40 cm thick in parts of Trench FA, but elsewhere around that trench and in Trench HH1, Y1 is quite thin to discontinuous (Figure 4.1, 4.2). The granulometric data (Figure 5.2) show a prominent spike of coarse rock fragments, especially in the 32–2 mm class, and only a small amount of clay-sized sediment. In addition, $CaCO_3$ has a pronounced peak in Y1. The rocks tend to be angular, and their porosity is quite low. These are not *éboulis secs*, however, because there is a modest amount of fine matrix; for example, compare the histograms of sediment samples FR 1–15 and 1–16 with the *éboulis secs* of FR 2–8 and 3–12 of Stratum V (Appendix C).

Stratum Y2 is 1 to 2 m thick in Trenches FA, HH1, and H2. It is difficult to identify in Trench GG1 because of deep disturbance there. This is a grayish brown loam of variable stoniness, including a number of head-sized rocks and occasional limestone blocks up to 80 cm across. This stratum is contemporaneous with the breakdown of the cave brow as clearly seen in Trench H2, where the lowest rocks of the breakdown sit on top of Unit H2B:31, roughly in the middle of Stratum Y2. Higher excavation units in Trench H2B lap against the breakdown pile. Scattered large blocks in other trenches may have fallen at the same time as the breakdown of the brow.

Stratum Y3 is distinguished only in Trench FA. Presumably it is mixed with the modern disturbance of Stratum Z in the other trenches. In FA this stratum reaches a maximum thickness of ca. 1 m. It contains a number of hearths, much ash and charcoal, some scattered head-sized (and larger) rocks, and shows ancient disturbance in the form of pits presumably dating to the Neolithic habitation, notably in the north part of Trench FAN and the adjacent FA East Balk.

There appears to be an unconformity separating Strata Y2 and Y3. The sedimentation rate in Y1 and Y2 is rather well constrained by a number of radiocarbon dates, although there is a scatter of dates presumably related to modern disturbance. There are only two dates for Stratum Y3, however, and, if taken at face value, they indicate a rapid rate of sedimentation. These two segments of the rate curve imply a hiatus of as much as 800 years, if the dates are valid. There is some support in the laboratory analyses for an unconformity at this point (Figure 5.2). The uppermost samples from Stratum Y2 show increased roundness, decreased $CaCO_3$, and some white rocks, but these changes do not seem to be of a magnitude that matches such a long hiatus. (Compare, for example, the changes associated with hiatuses of only 600 to 650 years following Strata V and X1.) Thus, this hiatus may not be as long as suggested on Figure 6.2. Its duration depends critically on a single date in Unit FAS:72, which may be spurious or in disturbed context. Nevertheless, I suggest that a short hiatus does occur here, or at least a strong reduction in the rate of sedimentation.

The human influence on sedimentation in the Y strata is conspicuous. First of all, the rate of sedimentation is comparable to that in Strata W2 and W3, which are heavily anthropogenic. There are numerous hearths and scatters of ash and charcoal. Ceramic artifacts are abundant from Stratum Y1 upward (Vitelli 1993). Sheep or goats appear in Y1 and are abundant thereafter, accounting for 80 to 90% of the megafauna in many units (Payne in Jacobsen 1973:63), and botanical remains increase to 1000 g or more of carbon per 100 years in biozone VIIb (Hansen 1991:Figure 61) for the first time since Stratum W3.

Modern Disturbance—Stratum Z

The top 1 to 3 m of the cave filling in the excavated areas is rather thoroughly disturbed. Wheel-thrown pottery, iron nails, bottle glass, and modern wood persist alongside Neolithic sherds at least as deep as −1.88 m below the surface in Trench H1 and H1 Terrace, and down to −2.23 m in Trench H, according to the excavators' notes. In addition, several fragments of copper were recovered at −2.82 m in Trench H1. Moreover, ash deposits occur throughout Stratum Z, two of which were radiocarbon-dated at A.D. 1845 and A.D. 1910.

The sediments themselves suggest disturbance. In part they are very loose and soft, as if they had been dumped recently (Plate 4b). In some areas they have a very homogeneous appearance without obvious stratification, especially in Trench HH1 and H1 Terrace. Moreover, they are rather rich in rock fragments and small boulders, as if the finer matrix had been selectively removed, suggesting exploitation for niter (saltpeter) or fertilizer. Looting may also have been a source of disturbance. For a more detailed discussion of disturbance, see Vitelli (1993:32–33), who was able to recognize the disturbance on the basis of ceramics recovered in unmixed contexts found in a few areas.

Overview of the Sedimentary Fill

Approximately 30,000 years of prehistory are represented in the excavated fill in Franchthi Cave, if one omits the post-Neolithic disturbance in the cave. However, well over half of that span of time is in the form of hiatuses, or times of nondeposition of sediment. On the other hand, more than 5 m of sediment older than ca. 40,000 B.P. underlies the excavated sequence in the front part of the cave, and a few artifacts from disturbed contexts suggest that Middle Palaeolithic occupations may be buried there.

The oldest dated event is the deposition of tephra blown in from the Naples area in Italy ca. 33,000 to 40,000 Cal B.P. After a probable hiatus just above the tephra, slow sedimentation of clay loams interspersed with large blocks of limestone fallen from the cave ceiling occurred. The cave was frequented by people who were hunting wild horses and cattle that may have been roaming on the broad coastal plain that separated the cave from the sea in those times. A couple of long hiatuses interrupted sediment accumulation in the Cave, especially a long interval from ca. 25,000 Cal B.P. until 15,000 Cal B.P., which happens to coincide with the maximum glacier expansion in central and northern Eurasia.

From 15,000 Cal B.P. onwards, Franchthi Cave was occupied more or less continuously except for several relatively brief hiatuses of 500 years or so each. The intensity of occupation varied with time, as can be seen in the sedimentation rate curve (Figure 6.2). In fact, there seems to have been three pulses of human activity in the cave separated by relatively quiet times. The first pulse saw the accumulation of prominent land-snail-shell middens in late Upper Palaeolithic time (Stratum T), followed by the clay loams of Stratum U with strongly reduced anthropogenic input, and ending with a prominent layer of coarse *éboulis secs*. After a hiatus of some 500 years, the cave was intensely occupied by Mesolithic peoples, who also left behind numerous snail-shell middens, as well as abundant evidence of plant collecting, hunting (mostly deer), and deep-sea fishing. In this second pulse the sedimentation rate in the cave increased abruptly to ca. 250 cm per 100 years for a period of a few centuries early in Mesolithic times.

The Mesolithic occupations ended with another layer of *éboulis secs*, followed by another 500-year hiatus. The early Neolithic utilization of the cave began quietly with relatively slow accumulation of grayish brown loam of Stratum X2, similar to that of Stratum U, which included sporadic appearances of ceramics and the appearance of domesticated plants in limited quantities, particularly emmer wheat. The third pulse begins with Stratum Y1, a rocky layer marking the introduction of domesticated sheep and/or goats in large numbers, abundant ceramics, and a large influx of domesticated plant remains—einkorn wheat, barley, oats, and more (Hansen 1991). This trend continues through Stratum Y2 and, after a possible short hiatus of a century or so, through Y3 until ca. 6000 Cal B.P. or later.

Rock falls from the cave ceiling occurred throughout, especially during the accumulation of the basal Strata P through S2, but again during the Neolithic (Stratum Y2 in Trench H2) and at the close of the Neolithic occupations, when the large window in the cave roof was opened by a major collapse.

APPENDIX A

List of Sediment Samples with Provenience, Size, Color, and Field Description

SAMPLE NUMBER	EXCAVATION UNITS	DEPTH TO BASE below datum (m)	THICKNESS cm	DRY WGT grams	MUNSELL SOIL COLORS moist	COLORS dry	FIELD DESCRIPTION based on DRY color
FR 1-1	FAN:70,74,79?	2.08	7	4800	10YR 2/2	10YR 2/6	Very dark brown stony loam with abundant charcoal and shells; sherds present
FR 1-2	FAN:81	2.17	6-7	4450	10YR 3/2	10YR 5/3	Brown stony loam; stones concentrated in lower part; stones and shells with chalky surfaces
FR 1-3	FAN:84,85	2.25	5-6	2150	10YR 2/1	2.5Y 4/2	Dark grayish brown loam with few stones; much charcoal; some bones and sherds
FR 1-4	FAN:89,90,94,95	2.45	17	4600	10YR 3/2	10YR 5/3	Brown loam with several large stones >128 mm; some sherds and charcoal
FR 1-5	FAN:108	2.62	8	2750	10YR 2/1	10YR 6/2	Light grayish brown loam with abundant charcoal; several angular blocks; some bones and sherds
FR 1-6	FAN:108, 110	2.78	17	4950	10YR 3/2	10YR 6/3	Pale brown loam, stonier but with less charcoal and bones than 1-5
FR 1-7	FAN:110,111	2.87	12	5100	10YR3/4	10YR 6/3	Pale brown loam with stones; some sherds, especially in lower part; surrounded by large rocks
FR 1-8	FAN:114,117,118	3.14	11	3400	10YR 3/4	10YR 5/4	Yellowish brown loam with few large stones; some sherds and shell
FR 1-9	FAN:121,122	3.38	10	6250	10YR 3/4	10YR 5/4	Yellowish brown stony loam with many sherds, and abundant charcoal
FR 1-10	FAN:127	3.55	10	3700	10YR 2/2	10YR 5/2	Grayish brown loam with very few stones; abundant charcoal fragments parallel to laminations; some sherds, marine shell
FR 1-11	FAN:133,134	3.72	7	2600	10YR 3/2	10YR 5/3	Brown loam with reddish brown streaks; few stones; some charcoal
FR 1-12	FAN:134 lower	3.85	12	4550	10YR 3/2	10YR 5/2	Grayish brown loam, like 1-11, but more stones; immediate adjacent to a reddened hearth
FR 1-13	FAN:141	4.04	9	4700	10YR 3/4	10YR 5/4	Yellowish brown compact loam with very few stones,and little cultural debris; some thin partings of light gray silt.
FR 1-14	FAN:146	4.15	11	4500	10YR 3/3	10YR 6/3	Pale brown loam with abundant stones; some charcoal, bones, etc.
FR 1-15	FAN:146 lower	4.17	4	3100	10YR 3/4	10YR 5/4	Yellowish brown sandy gravel; subangular stones mostly 2-3 cm; little culture
FR 1-16	FAN:148, 149?	4.23	7	2450	10YR 3/4	10YR 6/3	Pale brown sandy loam with very few stones, except in pockets
FR 1-17	FAN:148	4.30	7	3100	10YR 3/3	10YR 6/3	Pale brown silt loam, moist and compact; stones present but not abundant
FR 1-18	FAN:150,151	4.41	10	5700	10YR 3/3	10YR 6/3	Like 1-17, but stonier; only a scrap of bone
FR 1-19	FAN:153,156	4.55	10	4850	10YR 3/3	10YR6/3	Like 1-18, but less stony; very little cultural debris
FR 1-20	FAN:162,163	4.74	10	6900	10YR 3/3	10YR 5/3	Brown sandy loam; very abundant stones with punky, white surfaces; some bone, charcoal; occasional reddened stones
FR 1-21	FAN:165,167	4.91	12	9250	10YR 3/3	10YR 5/4	Yellowish brown slightly sandy loam with abundant largish, angular stones, and rounded pebbles
FR 1-22	FAN:172	5.10	7	5100	10YR 3/4	10YR 6/4	Light yellowish brown sandy loam like 1-21, but fewer, smaller, less angular stones
FR 1-23	FAN:172-174	5.19	10	6750	10YR 3/4	10YR 5/4	Yellowish brown fine-sandy, clayey, stony loam; very little matrix between stones
FR 1-24	FAN:179,180	5.31	10	5750	10YR 3/3	10YR 5/4	Lens of yellowish brown silty loam with moderately abundant stones; many snails; some charcoal and bones
FR 1-25	FAN:186 bottom	5.49	8	5750	10YR 3/4	10YR 5/6	Yellowish brown loam with moderate amount of stones; numerous rotten bone fragments; some snails
FR 1-26	FAN:191	5.69	10	4800	10YR 3/4	10YR 5/6	Yellowish brown loam with moderate amount of stones; much splintered bone; some snails
FR 1-27	FAN:195,196	5.84	9	5750	10YR 3/3	10YR 5/4	Like 1-26 but much less bone; many small snail shell fragments, especially at base of sample
FR 1-28	FAN:203,206	6.00	8	4500	10YR 2/2	10YR 5/4	Yellowish brown slightly sandy, stony clay loam; snails, bones, and a little charcoal; adjacent to a hearth
FR 1-29	FAN:215	6.23	8	3200	10YR 2/1	10YR 5/3	Brown clay loam with many snail fragments,but fewer than between 1-28 and 1-29
FR 1-30	FAN:218, 219 top	6.35	10	6800	10YR 2/1	10YR 5/3	Brown, somewhat gritty clay loam with many rocks of moderate size; occasional whole snail shells
FR 1-31	FAN:219 base, 221	6.47	9	6500	10YR 2/2	10YR 5/3	Brown clay loam like 1-30; stones punky with white rinds; some charcoal; appears to be in a pit, according to section drawing.
FR 1-32	FAN:222 lower,224	6.74	9	6500	10YR 2/1	10YR 5/3	Like 1-31, but much stonier; some shells and charcoal

SAMPLE NUMBER	EXCAVATION UNITS	DEPTH TO BASE below datum (m)	THICKNESS cm	DRY WGT grams	MUNSELL SOIL COLORS moist	dry	FIELD DESCRIPTION based on DRY color
FR 2-0	FAS:170,172	6.00	9	4500	10YR 3/4	10YR 6/4	Light yellowish brown (clay) loam, relatively few stones; some charcoal, snails, and flints; cf. FR-1-27; chalky limestones
FR 2-1	FAS:184,185 top	6.64	13	7500	10YR 3/2	10YR 5/3	Brown (clay) loam with abundant stones; some snails, bones; immediately below a charcoal lens in FAS:183
FR 2-2	FAS:185 base,186	6.81	10	7100	10YR2/1	10YR 5/3	Very similar to 2-1; some stones with thick white rinds and very fragile; some charcoal
FR 2-3	FAS:189,190	7.08	9	6400	10YR 3/2	10YR 5/3	Like 2-2; perhaps more charcoal; some bones
FR 2-4	FAS:191	7.24	7	5750	10YR 2/2	10YR 5/3	Brown very stony loam with many snail shells; some bone, marine shells, charcoal; overlies a mash of crushed snail shell
FR 2-5	FAS:192 upper	7.30	7	2200	10YR 2/2	10YR 5/3	Brown loam with very few stones; crushed shell abundant, but not as much as just above this sample
FR 2-6	FAS:195	7.55	20	7500	10YR 2/2	10YR 5/3	Brown very stony (clay) loam; abundant charcoal,snails, and bones
FR 2-7	FAS:196	7.69	8	5000	10YR 2/2	10YR 5/3	Like 2-6 but less stony; directly above a small hearth.
FR 2-8	FAS:199-201	7.95	14	6850	10YR 3/4	10YR 6/4	Light yellowish brown very stony (clay) loam; "redder" than overlying strata; stones with heavy white rinds; large charcoal fragments.
FR 2-9	FAS:203	8.06	9	6400	10YR 3/4	10YR 5/4	Yellowish brown (clay) loam, quite moist; scattered charcoal, and a pocket of snails
FR 2-10	FAS:204	8.19	6	3500	10YR 4/4	10YR 6/4	Light yellowish brown (clay) loam, a few stones; scattered charcoal, bone, dripstone fragments, rotten bone fragments
FR 2-11	FAS:205-207	8.43	12	6300	10YR 3/4	10YR 6.5/3	Pale brown clay loam like 2-10, but abrupt increase in stones; some bone, shell, charcoal.
FR 2-12	FAS:208	8.65	14	6250	7.5YR 4/4	7.5YR 6/4	Light brown clay loam with some stones, mostly small; note change to 7.5YR hues; quite a few snails; some charcoal, flint.
FR 2-13	FAS:209	8.94	15	7500	7.5YR 4/4	7.5YR 6/5	Light yellowish red clay with large and small stones; horse bones, some charcoal
FR 2-14	FAS:219	9.48	12	6000	7.5YR 4/4	7.5YR 5/6	Strong brown, crumbly , moist and sticky clay; few stones except around large boulders; a little bone.
FR 2-15	FAS:221 upper	9.72	9	4200	7.5YR 4/4	7.5YR 5/6	Strong brown sandy clay loam, weakly cemented; may contain reworked tephra
FR 2-16	FAS:221 lower	9.79	5-6	n.d.	n.d	10YR 6/2	Light brownish gray tephra; crusty (lightly cemented) in upper part; lies directly on large boulder.
FR 2-17	FAS:216,218,220	n.d.	5-10	5750	5YR 4/6	10YR 5/6	Yellowish brown gravelly clay loam; some bone; lies directly on tephra.
FR 2-18	FAS:221	n.d.	7	n.d.	n.d.	10YR 6/2	Light brownish gray tephra; sharp upper and lower contacts here; scattered rock fragments in upper part
FR 2-19	FAS:222	n.d.	2-10	n.d.	7.5YR 5/6	7.5 YR 6/6	Reddish yellow clay loam with very few stones, but pale yellow blotches (2-3 cm across) of decayed ls fragments; travertine crusts
FR 2-20	FAS:223	n.d.	irreg.	3650	7.5YR 4/6	7.5YR 5/6	Strong brown clay loam,soft and moist, with scattered medium-sized rock fragments with black (Mn?) stains
FR 3-1	H1B:108,109 top	4.22	8	4500	10YR 3/4	10YR 5/4	Yellowish brown, somewhat sandy loam with some stones, charcoal, snails, and bones
FR 3-2	H1B:109 lower,110	4.29-4.38	6	4350	10YR 3/4	10YR 5/4	Like 3-1
FR 3-3	H1B:115	4.38-4.48	7	7800	10YR 3/3	10YR 5/4	Yellowish brown éboulis secs; coarse angular stones with little matrix; abundant snails, charcoal, and flint some bones, including fish
FR 3-4	H1B:118,120	4.53-4/59	8	6750	10YR 3/3	10YR 5/4	Yellowish brown sandy loam, moderately stony, with charcoal, snails, some bone.
FR 3-5	H1B:124, 127 upper	4.79-4.86	8	5200	10YR 3/3	10YR 5/4	Yellowish brown somewhat sandy (clay) loam with charcoal, some bone and flint
FR 3-6	H1B:127 lower, 128	4.86-4.91	7	3600	10YR 2/2	10YR 5/3	Brown loam full of crushed snail shells and charcoal; some bone; many small flint chips
FR 3-7	H1B:130,132	4.99-5.05	7	4650	10YR 3/2	10YR 5/3	Brown sandy loam, rather loose; not many stones; some charcoal fragments and isolated whole snail shells
FR 3-8	H1B:134	5.16-5.19	9	6150	10YR 3/3	10YR 5/3	Brown loam with some rocks; abundant snails; some charcoal, flint
FR 3-9	H1B:138,139	5.39-5.46	9	4700	10YR 3/3	10YR 5/3	Like 3-8
FR 3-10	H1B:141 lower, 142	5.57-5.65	11	5600	10YR 3/3	10YR 6/3	Pale brown loam with moderate number of stones some scattered charcoal; little cultural debris
FR 3-11	H1B:144,145	5.83-5.86	12	5250	10YR 3/3	10YR 6.5/3	Pale brown loam, like 3-7 through 3-10 with minor differences in stoniness; moderate amount whole snails; some bone

SAMPLE NUMBER	EXCAVATION UNITS	DEPTH TO BASE below datum (m)	THICKNESS cm	DRY WGT grams	MUNSELL SOIL COLORS moist	dry	FIELD DESCRIPTION based on DRY color
FR 3-12	H1B:148,150	6.15	19	10,100	10YR 2/2	10YR 5/2	Grayish brown *éboulis secs*; almost no matrix; stained by ash from overlying hearth; abundant snails; stalactite fragments
FR 3-13	H1B:151, some152?	6.27	10	6850	10YR 3/3.5	10YR 6/3.5	Light yellowish brown, slightly clayey loam with abundant stones; more compact than above; conspicuous white rinds on stones
FR 3-14	H1B:155	6.49	8	4650	7.5YR 3/2	10YR 4.5/4	Yellowish brown (clay) loam with abundant snails and charcoal; a few small stones; some bone.
FR 3-15	H1B:156,157	6.58-6.68	7-15	4100	10YR 3/3	10YR 4.5/4	Yellowish brown loam with many bone fragments, some snails, which increase downward.
FR 4-1	H1B:158,159, 160?	7.02	14	5700	10YR 3/3	10YR 5/4	Yellowish brown loam with small to medium stones; some snails, bone, *Patella*; a little charcoal.
FR 4-2	H1B:168	7.49	6	4250	10YR 2/2	10YR 4/4	Dark yellowish brown loam, definitely different from usual sediment; few stones; almost no snails; little cultural debris.
FR 4-3	H1B:170,171	7.83	irreg	5400	10YR 3/2	10YR 4/4	Dark yellowish brown fine sandy loam with few, scattered stones; some (equid?) bone.
FR 4-4	H1B:172,173	8.10	irreg.	7500	7.5YR 5/6	7.5YR 6/6	Reddish yellow very stony clay loam with little, if any, cultural debris; fills holes and cracks in underlying rocks
FR 4-5	H1B:174-182	8.37	20	6250	7.5YR 5/6	7.5YR 7/4	Pink stony clay loam; one bone fragment noted.
FR 4-6	H1B:180-188	8.47	18	6250	7.5YR 5/6	7.5YR 6/6	Reddish yellow clay loam, like 4-5, but smaller rocks; contains bones, including equid.
FR 4-7	H1B:195-201 top	8.76	irreg.	4900	10YR 5/6	7.5YR 6/6	Reddish yellow clay loam, like 4-5 and 4-6; some bone fragments, marine shells, and microfauna
FR 4-8	H1B:195-201	8.72	17	3530	7.5YR 5/6	7.5YR 6/6	Like 4-7
FR 4-9	H1B:203?,204	9.00	16	3625	7.5YR 5/6	7.5YR 6/6	Reddish yellow clay loam with stone fragments; moderately abundant cultural debris
FR 4-10	H1B:206,207, 208?	9.17	17	3490	7.5YR 5/6	7.5YR 6/6	Like 4-9
FR 4-11	H1B:213	9.42	9	n.d.	7.5YR 5/6	7.5YR 6/6	Tephra mixed with yellowish red clay loam; partly cemented
FR 4-12	H1B:215	9.60	13	n.d.	7.5YR 5/6	7.5YR 6/6	Reddish yellow clay loam, cemented.

APPENDIX B

Analytical Data for Sediment Samples

See text for explanation of parameters.

Appendix B. Analytical Data for Sediment Samples—Page 1

SAMPLE FA	NORM. ALT. FA	SAMPLE HH1	NORM. ALT. HH1	128-32 FA	128-32 HH1	32-2 FA	32-2 HH1	2-0.063 FA	2-0.063 HH1	63-2 FA	63-2 HH1	< 2 µm FA	< 2 µm HH1	ROUNDNESS FA	ROUNDNESS HH1	POROSITY FA	POROSITY HH1
1-1	7.93			19		11		16.9		37.8		10.2		40.7		1.9	
1-2	7.83			34		16		12.8		25.3		10.3		51.8		1.6	
1-3	7.70			11		10		15.5		43.1		10.5		53.6		1.6	
1-4	7.41			20		9		18.1		35.1		19.2		33.6		1.7	
1-5	7.15			23		12		15.7		36.6		11.1		31.8			
1-6	7.11			35		11		11.4		29.2		12.4		50.1		0.9	
1-7	6.93			29		14		13.9		26.9		14.5		41.2		1.7	
1-8	6.75			7		20		18.6		33		18.6		44.1		1.7	
1-9	6.50			45		13		10.6		21.8		9.2		46.0		2.2	
1-10	6.29			6		12		17.7		44.2		18.9		37.3		3.2	
1-11	6.11			9		10		13.6		43.3		22.3		31.6		1.2	
1-12	5.99			14		15		14.2		39		16.7		35.3		2.3	
1-13	5.87			2		19		22.7		38.7		15.8		48.5		1.3	
1-14	5.80			11		34		20.1		22.3		11.3		34.5		0.8	
1-15	5.77			14		44		19.1		13.8		7.1		29.5		0.6	
1-16	5.72			6		24		20.8		31.4		16.8		26.5		0.7	
1-17	5.69			11		17		17.5		37.1		16.6		31.6		0.6	
1-18	5.55			39		10		9.6		26.2		14.3		29.1		0.6	
1-19	5.43	3-1	8.00	32	13	12	17	10.7	28.2	29	29.3	15.1	9.8	29.3	23.0		0.8
1-20	5.12	3-2	7.8	58	13	20	17	6.0	25.8	9.7	31.3	4.4	8.4	43.0	29.6		0.6
1-21	5.00	3-3	7.70	60	74	14	10	8.0	5.9	13	7.0	3.1	1.9	40.0	36.0		0.7
1-22	4.80	3-4	7.52	43	26	17	21	11.1	19.2	18.4	30.2	7.8	6.9	47.0	44.0		0.9
1-23	4.70	3-5	7.28	67	32	14	21	4.4	13.8	7.8	27.1	4.1	6.4	51.0	53.0		1.9
1-24	4.60	3-6	7.19	52	22	9	15	8.3	22.8	18.3	31.9	8.6	7.6	45.0	22.6		1.9
1-25	4.42	3-7	7.06	56	29	12	19	8.2	18.9	16.7	26.6	6.4	6.0	51.0	37.0		1.7
1-26	4.25	3-8	6.91	56	19	13	11	8.8	21.8	15.1	37.6	6.7	9.8	47.0	24.0		1.9
1-27	4.08	3-9	6.65	52	14	17	25	8.0	22.4	13.4	30.2	6.9	7.0	42.3	24.0	1.4	1.5
1-28	4.06	3-10	6.50	33	20	32	13	16.0	26.4	20.8	31.4	10.5	7.4	54.4	32.2	1.6	1.2
2-0	4.05	3-11	6.30	22	28	18	15	14.3	20.6	31.4	27.7	11.0	7.7	43.6	23.0	1.7	1.3
	3.67	3-12	6.00	40	80	32	11	10.3	2.2	12.6	4.9	3.8	0.6	38.7	37.0	1.7	1.4
	3.55	3-13	5.86	56	46	13	17	12.3	8.4	13	18.7	4.0	7.6	38.1	64.0	1.2	1.3
	3.43	3-14	5.60	41	16	22	40	13.9	13.6	17	19.2	5.4	9.9	44.0	68.0	2.3	1.6
	3.35	3-15	5.50	48	46	21	27	7.8	7.7	12.2	12.4	4.3	5.9	42.0	46.6	2.0	1.3
	3.20	4-1	5.40	56	42	22	34	9.4	9.6	9.6	9.0	2.8	3.6	44.0	45.6	2.6	0.7
	3.16	4-2	4.87	46	33	21	15	7.8	11.0	8.1	30.4	2.8	10.1	34.0	56.3	3.3	1.8
	2.90	4-3	4.60	48	42	25	8	13.4	10.1	13.6	26.5	4.6	11.5	35.0	63.0	2.9	2.3
	2.77	4-4	4.17	4	61	23	16	11.0	4.5	10.3	13.1	3.6	4.7	33.0	32.0	2.8	2.8
	2.70	4-5	3.92	54	26	22	28	32.2	19.0	27.5	19.3	9.9	6.9	n.d.	31.8	2.6	0.6
	2.46	4-6	3.80	31	37	24	19	9.9	13.4	10.8	19.4	3.0	7.9	41.5	26.6	2.4	1.6
2-7	2.30	4-7	3.57	57	16	20	21	16.5	18.1	19	12.6	5.2	12.6	37.4	52.1	1.4	1.1
2-8	2.03	4-8	3.52	60	15	18	21	7.7	16.3	12.4	14.7	3.7	14.7	47.0	34.2	2.1	1.4
2-9	1.90	4-9	3.27	7	33	11	20	8.1	14.6	10.9	8.9	4.4	8.9	49.6	53.2	2.4	1.1
2-10	1.79	4-10	3.09	24	11	32	25	24.8	28.7	39.3	5.8	17.8	5.8	n.d.	48.4	2.7	0.9
2-11	1.54	4-11	2.87	27		22		14.9		19.5		8.1		68.7		1.6	
2-12	1.31	4-12	2.69	50		24		13.3		23.7		12.2		55.0		0.9	
2-13	1.03							7.8		12.5		4.2		50.0		1.0	
2-14	0.50			33		14		11.3		27.2		13.8		30.0		0.8	
2-17	0.50			23		53		7.9		11.3		3.2		54.5		2.2	
2-20	0.01			13		7		16.1		42		21.6		61.1		9.5	

Appendix B. Analytical Data for Sediment Samples—Page 2

O.M.(LOI) FA	O.M. (LOI) HH1	CaCO3 (LOI) FA	CaCO3 (LOI) HH1	CaCO3 acid FA	CaCO3 acid HH1	pH FA	pH HH1	% CULTURE FA	% CULTURE HH1	SNAIL FRAGS 1-2 mm FA	SNAIL FRAGS 1-2 mm HH1	SAMPLE FA	SAMPLE HH1
8.7		36.4		56.1		7.85		3.90				1-1	
7.5		22.4		47.7		7.90		2.40		20.00		1-2	
10.0		29.6		40.7		7.70		7.60				1-3	
5.9		35.0		43.9		7.60		0.90		20.00		1-4	
7.6		34.6		42.4		7.70		2.00				1-5	
6.2		36.8		50.0		7.80		2.10		0.00		1-6	
4.3		37.8		48.2		7.80		2.10				1-7	
4.7		38.7		44.6		7.80		1.80		0.00		1-8	
4.7		35.4		42.1		7.80		5.30				1-9	
8.7		31.9		45.7		7.75		3.60		37.50		1-10	
6.7		35.3		49.2		7.90		1.60				1-11	
6.5		37.5		51.3		7.85		3.00		20.00		1-12	
4.6		38.1		50.4		7.90		1.20				1-13	
4.8		42.7		57.6		7.80		1.10		37.50		1-14	
2.8		56.0		63.5		8.00		0.70				1-15	
				55.4				0.20		20.00		1-16	
5.7		35.7		52.0		7.90		0.30				1-17	
4.6		37.4		55.9		7.90		0.10		20.00		1-18	
5.7	5.3	40.3	44.8	57.7	56.3	7.80	7.70	0.10	1.80	20.00		1-19	3-1
6.1	5.3	31.7	44.7	54.2	59.2	7.90	7.70	0.60	1.90	37.50		1-20	3-2
5.8	5.3	29.8	25.8	50.2	59.0	8.00	7.70	0.60	1.20			1-21	3-3
5.6		34.3		53.9		7.90		0.70		37.50		1-22	
6.2	6.7	18.3	44.2	43.9	62.8	7.80	7.80	0.70	3.20			1-23	3-4
7.0	8.0	23.7	32.3	50.0		7.75	7.75	3.80	1.90	37.50		1-24	3-5
6.1		22.3		43.8		7.60		5.00				1-25	
6.0		31.3		48.3		7.55		1.70		62.50		1-26	
5.8	11.4	34.3	41.5	49.4	58.6	7.70	7.70	1.70	5.20	87.50		1-27	3-6
6.9		30.0		41.7		7.75		3.00				1-28	
6.1	7.3	26.6	46.3	45.0	36.3	8.10	7.90	0.80	1.70			2-0	3-7
8.1		51.1		62.1		7.65		6.10				1-29	
6.6	7.5	59.2	45.1	66.7	58.9	7.85	8.05	0.90	3.10	87.50		1-30	3-8
6.9		57.9		60.1		7.80		1.10		62.50		1-31	
6.3		59.8		69.5		8.00		1.50				2-1	
7.8	6.7	56.2	46.6	67.0	42.6	7.80	7.80	0.90	2.90	62.50		1-32	3-9
8.7		56.8		65.5		8.00		1.10		50.00		2-2	
9.2		50.5		69.8		8.15		2.20		62.50		2-3	
6.8	6.3	61.7	49.7	68.2	61.8	8.00	8.05	3.70	1.30	50-60	62.50	2-4	3-10
8.6	5.2	54.4	51.6	69.4	61.8	8.10		6.90	2.30		62.50	2-5	3-11
6.6	12.0	58.3	35.1	69.0	54.9	8.05	7.85	1.40	0.90	50.00	37.50	2-6	3-12
6.8		53.3		66.2		8.15		4.10				2-7	
5.9	7.2	33.3	13.5	55.3	44.3	8.00	7.90	0.30	0.50	20.00	37.50	2-8	3-13
5.6		28.8		52.3		7.80		0.10				2-9	
5.0		16.1		40.8		7.80		0.30		< 10		2-10	
5.1		30.3		50.3		7.65		1.30				2-11	
	7.0		28.0		41.6		7.85		6.60		87.50		3-14
	9.2		20.7		52.6		7.90		2.30		62.50		3-15
6.3	6.1	16.6	36.4	40.5	50.5	7.85	7.90	1.40	2.90	10.00	37.50	2-12	4-1
	9.8		6.2		43.6		7.90		2.80		20.00		4-2
	7.5		5.4		44.6		8.00		2.10		20.00		4-3
	4.8		30.8		45.4		8.05		0.01		5.00		4-4
	3.2		46.0		59.0		8.20		0.90				4-5
4.6	4.8	30.5	32.4	53.9	51.7	7.90	8.15	2.20	3.90			2-13	4-6
	5.3		41.5		56.3		8.10		1.80				4-7
	5.7		27.3		36.9		7.75		0.20				4-8
	4.4		36.7		53.6		8.10		2.70				4-9
	3.7		38.2		58.2		8.10		0.30				4-10
					65.9								4-11
4.3		13.5		32.4	85.7	7.90		0.40		tr		2-14	4-12
4.4		16.1		35.0		8.20		0.30				2-17	
4.8		5.1		34.9		7.80		0.06		5.00		2-20	

APPENDIX C

Granulometric Histograms for All Sediment Samples Arranged by Stratum

Correlative samples from Trenches FA and HH1 are placed side by side for comparison. In a few cases, two histograms are superposed when it was possible to trace one sample physically to an adjacent one in the same trench complex. To determine which excavation units are included in each sediment cample, see Appendix A.

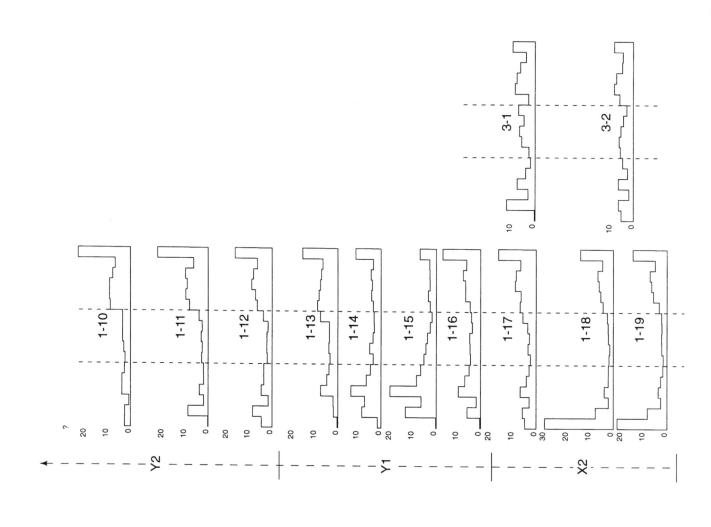

FRANCHTHI
Total Granulometry
(not acid-treated)

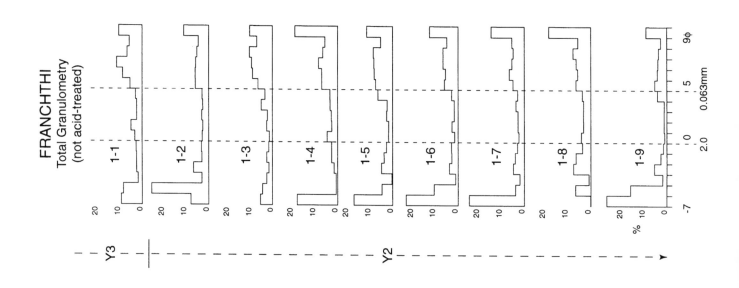

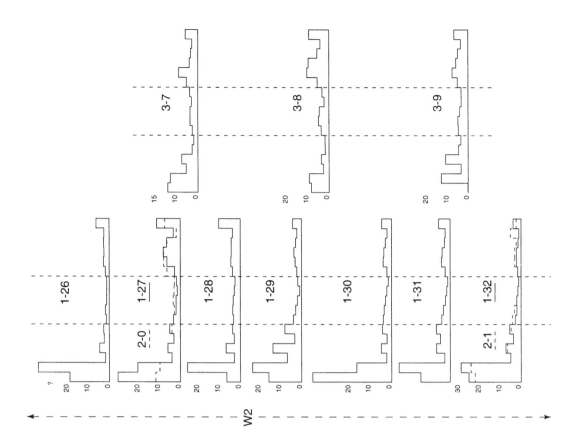

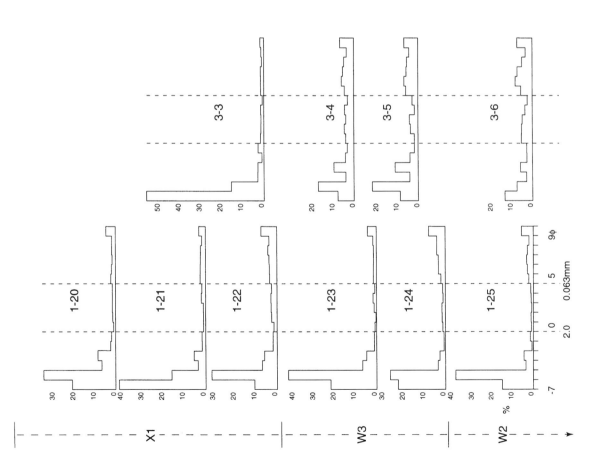

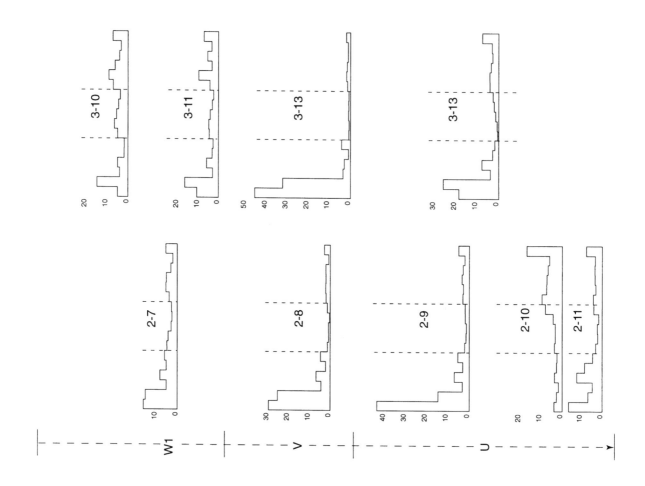

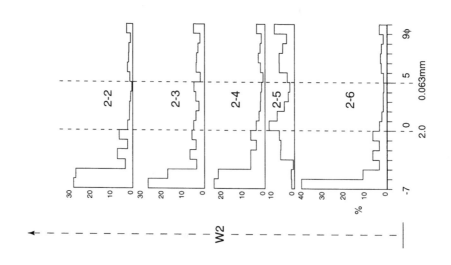

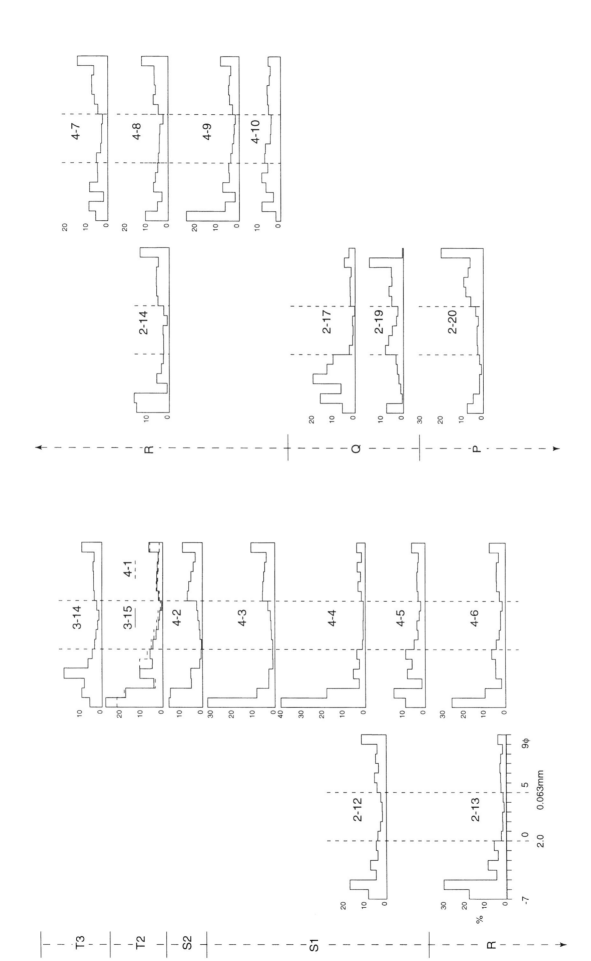

REFERENCES CITED

Andonakatou, D.
  1973   *Argolidhos Periiyisis*. Nomarkhias Argolidhos, Navplion, Greece.
Army Map Service
  1954   Sheet #2015 IV. Corps of Engineers, U. S. Army.
Bard, E., M. Arnold, R. G. Fairbanks, and B. Hamelin
  1993 ^{230}Th-^{234}U and ^{14}C ages obtained by mass spectrometry on corals. Radiocarbon 35(1):191–199.
Bard, E., B. Hamelin, R.G. Fairbanks, and A. Zindler
  1990   Calibration of the 14C Timescale over the Past 30,000 Years Using Mass Spectrometric U-Th Ages from Barbados Corals. *Nature* 345(62/4):405–410.
Binford, L. R.
  1968   *Post-Pleistocene Adaptations, New Perspectives in Archaeology*. Aldine, Chicago.
Braidwood, R. J.
  1960   The Agricultural Revolution. *Scientific American* 203:130–148.
Braidwood, R. J., and B. Howe
  1960   *Prehistoric Investigations in Iraqi Kurdistan*. Studies in Ancient Oriental Civilization Vol. 31, University of Chicago Press, Chicago.
Bursian, C.
  1872   *Geographie von Griechenland*. B. G. Teubner, Leipzig, Germany.
Butzer, K. W.
  1982   *Archeology as Human Evolution*. Cambridge University Press, Cambridge, UK.
Clark, J. G. D.
  1952   *Prehistoric Europe, The Economic Basis*. Methuen, London.
  1957   *Archaeology and Society*. Methuen, London.
CLIMAP Project members
  1976   The Surface of the Ice-Age Earth. *Science* 191:1131–1137.
Cornell, W., S. Carey, and H. Sigurdsson
  1983   Computer Simulation of Transport and Deposition of the Campanian Y-5 Ash. In *Explosive Volcanism*, Elsevier, Amsterdam, pp. 89–109.
Cullen, T.
  1995   Mesolithic Mortuary Ritual at Franchthi Cave, Greece. *Antiquity* 69(263):270–289.
Dakaris, S. I., E. S. Higgs, and R. W. Hey
  1964   The Climate, Environment and Industries of Stone Age Greece, Part I. *Proceedings of the Prehistoric Society* 30:199–244.
Diamant, S.
  1979   A Short History of Archaeological Sieving at Franchthi Cave, Greece. *Journal of Field Archaeology* 6:203–217.
Efstathiadi Group
  1970   *Peloponnese: Tourist Map and Guide*. P. Efstatiadis and Sons, Athens, Greece.
Farrand, W. R.
  1975   Analysis of a Prehistoric Rockshelter: the Abri Pataud. *Quaternary Research* 5:1–26.
  1979   Chronology and Palaeoenvironment of Levantine Prehistoric Sites as Seen from Sediment Studies. *Journal of Archaeological Science* 6:369–392.
  1985   Rockshelter and Cave Sediments. In *Archaeological Sediments in Context*, edited by J. K.

Stein and W. R. Farrand, pp. 21–39. Center for the Study of Early Man, University of Maine, Orono.

1988   Integration of Late Quaternary Climatic Records from France and Greece. In *Upper Pleistocene Prehistory of Western Eurasia*, edited by H. L. Dibble and A. Montet-White, pp. 305–319. University Museum Monograph 54, University of Pennsylvania, Philadelphia.

1993   Discontinuity in the Stratigraphic Record: Snapshots from Franchthi Cave. In *Formation Processes In Archaeological Context*, edited by P. Goldberg, D. T. Nash, and M. D. Petraglia, pp. 85–96. Monographs in World Archaeology 17, Prehistory Press, Madison, Wisconsin.

Geological Society of America

1991   *Rock-Color Chart*. The Geological Society of America, Boulder, Colorado.

Gibbon, G.

1984   *Anthropological Archaeology*. Columbia University Press, New York.

1985   Classical and Anthropological Archaeology: A Coming Rapprochement. In *Contributions to Aegean Archaeology: Studies in Honor of William A. McDonald,* edited by N. C. Wilkie and W. D. E. Coulson, pp. 283–294. University of Minnesota Press, Minneapolis.

Gifford, J. A.

1990   Analysis of Marine Sediments off Franchthi Cave. In *Franchthi Paralia: the Sediments, Stratigraphy, and Offshore Investigations*, by T. J. Wilkinson and S. T. Duhon, pp. 85–116. Excavations at Franchthi Cave, Greece, fasc. 6, Indiana University Press, Bloomington and Indianapolis.

Gillot, P-Y., S. Chiesa, S., G. Pasquaré., and L. Vezzoli

1982   <33,000–yr K-Ar Dating of the Volcano-Tectonic Horst of the Isle of Ischia, Gulf of Naples. *Nature* 299:242–245.

Goldberg, P. S., and Y. Nathan

1975   The Phosphate Mineralogy of et-Tabun Cave, Mount Carmel, Israel. *Mineralogical Magazine* 40:253–258.

Guillien, Y., and J. Lautridou

1970   Recherches de Gélifraction expérimentale du Centre de Géomorphologie. I-Calcaires des Charentes. *Bulletin du Centre de Géomorphologie de Caën* 5:1–45.

Hallerstein, H. v.

1986   *Streifzuge auf dem Peloponnes*. Otto Muller Verlag, Salzburg, Austria.

Hansen, J. M.

1991   *The Palaeoethnobotany of Franchthi Cave*. Excavations at Franchthi Cave, Greece, fasc. 7, Indiana University Press, Bloomington and Indianapolis.

Higgs, E. S., and C. Vita-Finzi

1966   The Climate, Environment, and Industries of Stone Age Greece, Part III. *Proceedings of the Prehistoric Society* 33:1–29.

Hole, F., K. V. Flannery, and J. A. Neely

1969   *Prehistory and Human Ecology of the Deh Luran Plain*. University of Michigan Press, Ann Arbor, Michigan.

Jacobsen, T. W.

1969a   Excavations at Porto Cheli and Vicinity. Preliminary Report, II: The Franchthi Cave, 1967–1968. *Hesperia* 38:343–381.

1969b   The Franchthi Cave, a Stone Age Site in Southern Greece. *Archaeology* 22:4–9.

1973   Excavation in the Franchthi Cave, 1969–1971. *Hesperia* 42:45–88.

1979   Excavations at Franchthi Cave. *Arkhaiologikon Dheltion, Khronika* 29:268–282.

Jacobsen T. W., and W. R. Farrand

1987   *Franchthi Cave and Paralia. Maps, Plans, and Sections*. Excavations at Franchthi Cave, Greece, fasc. 1, Indiana University Press, Bloomington and Indianapolis.

Jacobsen, T. W., and J. S. Kopper

1981   Excavations at the Frachti [sic] Cave in the Argolis. *International Speleological Meeting, Athens 1971*. Greek Speleological Society, Athens, pp. 80–105.

Jameson, M. H.
    1953   Inscriptions of the Peloponnesos. *Hesperia* 22:148–171.
    1959   Inscriptions of Hermione, Hydra, and Kasos. *Hesperia* 28:109–120.
Jameson, M. H., C. N. Runnels, and Tj. H. van Andel
    1994   *A Greek Countryside. The Southern Argolid from Prehistory to the Present Day.* Stanford University Press, Stanford, California.
Kromer, B. and B. Becker
    1993   German Oak and Pine 14C Calibration, 7200–9400 B. C. *Radiocarbon* 35(1):125–135.
Kyrou, A. K.
    1990   *Sto Stavrodhromi tou Argolikou,* vol. 1. Privately published, Athens, Greece.
Laville, H., J. Rigaud, and J. Sackett
    1980   *Rockshelters of the Perigord.* Academic Press, New York.
Leake, W. M.
    1846   *Peloponnesiaca: a Supplement to Travels in the Morea.* J. Rodwell, London.
    1830   *Travels in the Morea.* J. Murray, London.
McBurney, C. B. M.
    1967   *The Haua Fteah (Cyreniaca) and the Stone Age of the South-East Mediterranean.* Cambridge University Press, Cambridge, UK.
McDonald, W. A., and G. J. Rapp
    1972   *The Minnesota Messenia Expedition, Reconstructing a Bronze Age Regional Environment.* University of Minnesota Press, Minneapolis.
Mellaart, J.
    1967   *Çatal Hüyük. A Neolithic Town in Anatolia.* Thames and Hudson, London.
Meritt, L. S.
    1984   *History of the American School of Classical Studies in Athens, 1939–1980.* American School of Classical Studies, Princeton, NJ.
Meyer, E.
    1930   Mases. In *Real Encyclopädie der classischen Alterstumwissenschaft* 14. J. B. Metzler, Stuttgart, Germany.
Miliarakis, A.
    1886   *Yeografiki politiki nea kai arkhaia tou nomou argolidhos kai Korinthias.* Athens, Greece.
Munsell Color
    1975   *Munsell Soil Color Charts.* Munsell Color, Baltimore, MD.
Mylonas, G. E.
    1928   *I Neolithiki Epokhi en Helladhi.* Athens, Greece.
Paterne, M., F. Guichard, F., J. Labeyrie, P-Y. Gillot, and J-C. Duplessy
    1986   Tyrrhenian Sea Tephrochronology of the Oxygen Isotope Record for the Past 60,000 Years. *Marine Geology* 72:259–285.
Payne, S.
    1982   Faunal Evidence for Environmental/Climatic Change at Franchthi Cave (Southern Argolid, Greece), 25,000 B.P.–5,000 B.P.: Preliminary Results (abstract). In *Palaeoclimates, Palaeoenvironments, and Human Communities in the Eastern Mediterranean Region in Later Prehistory,* edited by J. L. Bintliff and W. van Zeist. British Archaeological Reports, International Series 133:133–137.
    1985   Zoo-Archaeology in Greece: a Reader's Guide. In *Contributions to Aegean Archaeology: Studies in Honor of William A. McDonald,* edited by N. C. Wilkie and W. D. E. Coulson, pp. 211–244. University of Minnesota Press, Minneapolis.
Pearson, G. W., B. Becker, and F. Qua
    1993   High-Precision Measurement of German and Irish Oaks to Show the Natural ^{14}C Variations from 7890 to 5000 B.C. *Radiocarbon* 35(1)93–104.
Perlès, C.
    1987   *Les Industries Lithiques Taillées de Franchthi (Argolide, Grèce).* Excavations at Franchthi

Cave, Greece, fasc. 3, Indiana University Press, Bloomington and Indianapolis.

1990 *Les Industries Lithiques Taillées de Franchthi (Argolide, Grèce).* Excavations at Franchthi Cave, Greece, fasc. 5, Indiana University Press, Bloomington and Indianapolis.

Pope, K. O., and Tj. H. van Andel
1984 Late Quaternary Alluviation and Soil Formation in the Southern Argolid: Its History, Causes, and Archaeological Implications. *Journal of Archaeological Science* 11:281–306.

Pouqueville, F. C. H.
1827 *Voyage de la Grèce.* 2d ed., Paris, France.

Renfrew, C.
1972 *The Emergence of Civilisation.* Methuen, London.
1980 The Great Tradition Versus the Great Divide: Archaeology as Anthropology. *American Journal of Archaeology* 84:287–298.
1983 Divided We Stand: Aspects of Archaeology and Information. *American Antiquity* 48(1):3–16.

Royal Hellenic Navy
1958 *Argolikos Kolpos, Sea of Hermione, Kolpos Methanon.* Greek Chart no. 64, Hydrographic Office of the Royal Hellenic Navy, Athens, Greece.

Shackleton, J.
1988 *Marine Molluscan Remains from Franchthi Cave.* Excavations at Franchthi Cave, Greece, fasc. 4: Indiana University Press, Bloomington and Indianapolis.

Spiliopoulou, T. I.
1965 I Spilia tou Kyklopa. *Periiyitiki* 83:28.

St. Seymour, K., and K. Christanis.
1995 Correlation of a Tephra Layer in Western Greece with a Late Pleistocene Eruption in the Campanian Province of Italy. *Quaternary Research* 43:46–54.

Stuiver, M. and G. W. Pearson
1993 High-Precision Bidecadal Calibration of the Radiocarbon Time Scale, AD 1950–500 BC and 2500–6000 BC. *Radiocarbon* 35(1)1–23.

Thunell, R., A. Federman, S. Sparks, and others
1979 The Age, Origin, and Volcanological Significance of the Y-5 Ash Layer in the Mediterranean. *Quaternary Research* 12:241–253.

Topping, P.
1976 Premodern Peloponnesus: The Land and People under Venetian Rule (1685–1715). *Annals of the New York Academy of Sciences* 268:92–108.

Trigger, B.
1986 *Prehistoric Archaeology and American Society, American Archaeology Past and Future.* Smithsonian Institution Press, Washington, D. C.

United States Office of Geography
1960 *Greece. Official Standard Names Approved by the U. S. Board on Geographic Names.* Office of Geography, Department of Interior, Washington, DC.

van Andel, Tj. H.
1998a Middle and Palaeolithic Environments and the Calibration of 14C dates beyond 10,000 BP. *Antiquity* 72:26–33.
1998b Paleosols, Red Sediments, and the Old Stone Age in Greece. *Geoarchaeology* 13(4):361–190.

van Andel, Tj. H., T. W. Jacobsen, J. B. Jolly, and others
1980 Late Quaternary History of the Coastal Zone near Franchthi Cave, Southern Argolid, Greece. *Journal of Field Archaeology* 7(4):389–402.

van Andel, Tj. H., and S. B. Sutton
1987 *Landscape and People of the Franchthi Region.* Excavations at Franchthi Cave, Greece, fasc. 2, Indiana University Press, Bloomington and Indianapolis.

Vitaliano, C. J., S. R. Taylor, W. R. Farrand, and T. W. Jacobsen

1981 Tephra Layer in Franchthi Cave, Peloponnesos, Greece. In *Tephra Studies as a Tool in Quaternary Research*, Proceedings of the NATO Advanced Study Institute, edited by S. Self and R. S. J. Sparks, pp. 373–379. Reidel, Dordrecht, Netherlands.

Vitelli, K. D.

1993 *Franchthi Neolithic Pottery, volume 1*. Excavations at Franchthi Cave, Greece, fasc. 8, Indiana University Press, Bloomington and Indianapolis.

1999 *Franchthi Neolithic Pottery, Volume 2: The Later Neolithic Ceramic Phases 3 to 5*. Excavations at Franchthi Cave, Greece, fasc. 10, Indiana University Press, Bloomington and Indianapolis, in press.

Weinberg, S. S.

1965 *The Stone Age in the Aegean*. The Cambridge Ancient History, fasc. 36, revised, pp. 3–68; also In Cambridge Ancient History 3d. ed., vol. 1, pt. 1, pp. 557–618 (1970). Cambridge University Press, Cambridge, UK.

Whitney-Desautels, N.

Forthcoming *The Freshwater and Land Snails from Franchthi Cave*. Excavations at Franchthi Cave, Greece, fasc. 11, Indiana University Press, Bloomington and Indianapolis.

Plates

Plate 1a. View of the Franchthi headland looking NE across Kiladha Bay. The entrance to Franchthi Cave and both breakdown windows can be seen in left center. The village of Kiladha is at the foot of the slope in the foreground.

Plate 1b. View of the Franchthi headland from the middle of Kiladha Bay. The cave mouth is at the left, and the middle breakdown window is in the middle of the scene.

Plate 2b. Cave floor just inside the entrance as seen from the top of the breakdown pile. Trench HH1 is prominent in the middle of the scene, and Trench FA appears on the right edge. The middle window breakdown can be seen in the background.

Plate 2a. Entrance to Franchthi Cave. Note the near-vertical stratification in the bedrook above the entrance and the pile of breakdown rubble under the brow of the cave.

Plate 3a. View of the presently habitable floor of Franchthi Cave seen from the top of the middle breakdown. Trench HH1 is just right of center, and H2 can be seen at the foot of the breakdown under the brow. Trench GG1 is on the left of the path into the cave, and FA is at left center surrounded by large piles of rocks removed from that trench.

Plate 3b. View from the top of the big breakdown pile under the middle window toward the rear of the cave where daylight is coming through the rear window. Note stalactites on the ceiling, and dripstone mounds on the fallen blocks.

Plate 4b. Uppermost, loose, and mostly reworked cave sediment at the top of the south end of Trench HH1. White ash deposits near the top are modern in age.

Plate 4a. The large chasm in the midst of the gigantic breakdown blocks under the middle window. This is location "1979" in Figure 2.1, also shown schematically in Figure 3.2. The top of the chasm is about 2–3 m wide as seen here.

Plate 5b. Land snail midden in Unit H1A:175 on the west face of Trench H1A. Each section on the folding scale is about 20 cm. Note abundance of charcoal. A second midden is seen at the bottom.

Plate 5a. The middle section of Trench H1B South, from approximately Unit H1B:115 to Unit H1B:160 (on the top of the large blocks at the bottom of this view). The one-meter scale leans against the column from which sediment samples FR 3-3 through FR 3-15 were removed.

Plate 6b. Detail of the bedrock wall at the far inner end of Franchthi Cave. Although the bedrock is intact, the highly fractured nature of the limestone can be seen clearly. Compare the fragment sizes to those in Plate 6a. Each section of the folding scale is about 20 cm.

Plate 6a. *Eboulis secs* (openwork rubble) in Stratum V (Units H1A:160 through H1A:165) on the north face of Trench HH1. Note the angularity of the rock fragments and the sparse matrix between them in contrast to the abundance of fine matrix above and below the *éboulis secs*. Each section of the folding scale is about 20 cm.

Plate 7a. Upper part of the west face of Trench FA. Trench A is to the left of the man. The section shown here is all Neolithic. Note the abundance of hearths, especially in Trench A.

Plate 7b. Detail of the pale gray volcanic tephra Stratum Q (Unit FAS:221). Note the sharp lower contact and the diffuse upper contact. The tephra is about 5-6 cm thick here. The light-colored spots, especially in Stratum P below the tephra, are decayed bone and fragments of limestone. These layers are draped over large limestone blocks as can be seen in the lower right corner.

Plate 8b. Middle part of the sediment sampling column on the west face of Trench FAN. These layers contain mostly early Neolithic artifacts. Note overlap of this view with Plate 8a.

Plate 8a. Upper part of the sediment sampling column on the west face of Trench FAN. These are upper Neolithic strata; Unit FAN:99 occurs about halfway down this part of the column. The width of the column is about 25-30 cm. Note the larger blocks in the lower part of the column, which can seen also in Plate 8b.

Plate 8d. Stony Mesolithic strata seen tangentially along the east face of Trench FA, looking south. Stratum X1 (*éboulis secs*) dips to the south away from the viewer

Plate 8c. The Neolithic/Mesolithic boundary in Trench FAN occurs at the abrupt shift from darker and moister (below) to lighter and drier color (above). Unit FAN:161 is situated just above the boundary. Stratum X1 (*éboulis secs*, Units FAN:163, 164) occurs immediately below the boundary. The large limestone block at the top of this view also appears at the bottom of Plate 8b.

INDEX